Springer Geophysics

The Springer Geophysics series seeks to publish a broad portfolio of scientific books, aiming at researchers, students, and everyone interested in geophysics. The series includes peer-reviewed monographs, edited volumes, textbooks, and conference proceedings. It covers the entire research area including, but not limited to, applied geophysics, computational geophysics, electrical and electromagnetic geophysics, geodesy, geodynamics, geomagnetism, gravity, lithosphere research, paleomagnetism, planetology, tectonophysics, thermal geophysics, and seismology.

More information about this series at http://www.springer.com/series/10173

Yuri D. Kononov

Long-term Modeled Projections of the Energy Sector

An Incremental Approach to Narrowing Down the Uncertainty Range

With a Foreword by Alexey A. Makarov
Translated by Svetlana V. Steklova

 Springer

Yuri D. Kononov
Irkutsk, Russia

ISSN 2364-9119   ISSN 2364-9127   (electronic)
Springer Geophysics
ISBN 978-3-030-30535-2   ISBN 978-3-030-30533-8   (eBook)
https://doi.org/10.1007/978-3-030-30533-8

This Springer imprint is published by the registered company Springer Nature Switzerland AG
The registered company address is: Gewerbestrasse 11, 6330 Cham, Switzerland

The book while presenting the research findings of the recent years is essentially a compendium of the research that has been carried out by the author over the span of several decades. Thus, I find it appropriate to have this book dedicated, in deepest affection and gratitude, to the memory of my teachers (first and foremost, to Lev A. Melentiev and Aleksandr S. Nekrasov) and my beloved late parents.

Russian Edition Editor's Foreword

The response of liberal-minded economists to the collapse of the Soviet planned system and the intensified stresses of the new age was not only to have little interest in but also to foster an ironic attitude toward forecasting of the development of complex processes in general and large industrial and economic systems in particular. Recently, yet another freshly minted B.Sc. graduate of an American university that ranks not far below the top uttered the following: "forecasting is the least intellectual of all activities," and then indulged in his not-exactly-original ideas of the present and the future.

In the meantime, it is exactly the widening of the uncertainty area under the rapid growth of production facility concentration, project costs, and delivery times that invoke the urgent need of project developers and, all the more so, project investors to have a better overview of prospective needs, opportunities, and risks. To satisfy this need, it is important not only to have an adequately constructed vision of the future but also to have an opportunity to discuss in tangible terms its alternatives and to have a platform for quantitative assessment of threats and arriving at trade-offs for the parties (stakeholders) that make decisions. A prerequisite of their interest in such a platform would be their trust in the methodology behind projections. The latter has to provide a sufficiently comprehensive formalized description of the premises and tools, and, most importantly, to independently reproduce the answers to the posed questions.

For the purposes of the energy sector development projections, such methodology, as applicable to the energy systems studies, began to develop almost half a century ago at what is now the Melentiev Energy Systems Institute of the Siberian Branch of the Russian Academy of Sciences, with the active participation of both the author and the editor of this book. Since then, it has gained international acceptance, having been improved upon and later adjusted in the course of the difficult transition from the central planning to a market-driven economic system. Nowadays, in Russia and worldwide, it is implemented in computer networks through complex systems of mathematical models and huge distributed databases.

As of today, worldwide and in Russia alike, there are several relatively comprehensive modifications of such model and information systems. In most cases, they have been created by skilled teams of specialists, they undergo the process of continuous improvement, and for many years they have been applied by them in the interactive mode for the purposes of projections of the world, national, and regional energy systems, the industries, and subsystems of the energy sector. The most advanced of the model systems are routinely employed for "rolling planning" of energy systems that are subjected to the regular analysis of their composition and root causes of the discrepancies between actual values and their previously published projections.

The key defining feature of modern methods of formalizing the forecasting process with respect to such complex production and economic systems as is the energy sector is matching the knowledge and intuitions of the experts skilled in this subject area with formalized tools of the input information processing and mathematical modeling of the development of the system that is subjected to forecast analysis. A symbiosis of this kind is meant for attaining increasingly deeper understanding of the objective trends and, if we are lucky enough, the laws governing its development. Philosophy-wise, this is an endless creative process that at no moment can guarantee the certainty of the projection one has arrived at, but (after a certain period of time) enables to identify the errors and/or incomplete knowledge of the system ex-post, so as to adjust the tools in search of the ways to eliminate them. The bottom line is the continuous enhancement of the platform that is meant to host focused discussions held by stakeholders with respect to costly projects, and, more importantly, to assess and distribute between them the risks of their implementation.

It is in this light that I've perceived and highly recommend this monograph by Prof. Dr. Yuri D. Kononov, one of the best experts in the world who specializes in one of the most complex areas, that of the study of external links of the energy sector. The above problem was formulated by him back in the 1970s and was later subjected to consistent and meticulous development at the Melentiev Energy Systems Institute of the Siberian Branch of the Russian Academy of Sciences and the International Institute for Applied Systems Analysis (Laxenburg, Austria). It has been drastically extended and is treated in the book in an inspired way (that is to say comprehensive, accurate, and engaging) covering various kinds of interactions with the external environment relevant to the energy sector that serves as the driving force and the giver sector of the Russian economy of the present (which will remain unchallenged in the nearest future).

It is exactly these "domestic" external factors of the energy sector, together with the links to the global energy markets that are very strong in the case of Russia, that account for the key contingencies and hence the uncertainty of projections. The monograph is dedicated to the discussion of the ways that are aimed at significant mitigation of their negative impact even if they do not completely rule out such impact.

The author begins the book by demonstrating the dependencye of the growth of the range of projection errors on the planning horizon (the demonstration is based, however, only on the highly aggregated indicators of the energy systems development) and then proposes and details an approach to the assessment of the pattern of losses for the project investor over time as dependent on the project delivery time and projection uncertainty, which is illustrated by select case studies. That said, the problem of the acceptable value of the error remains open for discussion. This is followed by an overview of inherent limitations and heuristic techniques with respect to the decrease of the impact of uncertainty it exerts on the decisions that have to be made (including the scenario approach), which is of much use, and there is a well-structured and concise yet reasonably comprehensive review of the formal methods employed to "combat" various types of uncertainties.

Having outlined the evolution undergone by Russian and international modeling tools employed for the forecasting of the energy industry development, the author proposes his three-step multistage procedure for narrowing down the projection range as applied to the process of making energy sector development projections. It supplements the recently enacted Russian law on strategic planning (the law covers the 15–20-year time frame) with the long-term projection step for the time frame that goes beyond 25 years into the future and, which is of much importance, includes approaches to the identification of the projection range boundaries and frequency-based ranking of facilities that fall within the range.

Unfortunately, the experience accumulated over the span of forty years while making projections of the long-term development of the USSR energy sector attests to the fact that the increase in the "farsightedness" and the level of elaboration of a long-term projection is oftentimes hollowed out by major bifurcations that take place as soon as in the first years of its actual implementation. To illustrate, the "Perestroika" launched by M. Gorbachev and the subsequent disintegration of the country voided "the long-term USSR energy program" developed in the early 1980s, while the 2008 crisis made "the energy strategy of the Russian Federation to year 2030" useless.

The subsequent chapters systematically cover the key issues of the energy sector development projections: they are methods of calculating fuel and energy demand and prices together with their interdependence (i.e., elasticity), the studies of the development inertia of energy systems, the barriers and risks that arise when overcoming the inertia, as well as methods and indicators that serve for their quantitative assessment. I believe that the reader will be interested in delving into the details of the approaches and findings obtained by the author on their own so as to make my commentary on them superfluous.

The book provides a floor for a concise, well-written (which is quite rare in the field), and well-informed (by Russian and international published research alike) discussion of the cutting edge of the complex and highly relevant issues of long-term energy sector projections, which is presented in a professional and intelligent way. Alongside the main target group made up of researchers and experts on the energy industry development, the book will be undoubtedly instrumental as an essential reading for master and Ph.D. students who major in "energy systems studies" and "complex problems of the energy systems development."

Moscow, Russia   Alexey A. Makarov
Academician

Presidium Member of the Russian
Academy of Sciences

Preface

Long-term projections serve as the initial stage of any analytical study of energy systems development and back up the conjectures regarding the development prospects. They are meant to outline the space of possibilities for feasible and efficient development of the national energy sector, to identify matters of concern and the bottlenecks to be addressed while pursuing such development, to set development targets alongside providing the foundation and data required to further and detail the analysis as part of the development of the energy strategy and policy, general schemes (master plans), and programs of the development of energy systems of individual industries and regions, as well as strategic plans of energy companies (see Table 1). Long-term projections are also crucial for laying the forward-looking groundwork in a broad area of knowledge related to the energy industry development.

The recently enacted law "On Strategic Planning in the Russian Federation" [1] makes energy industry projections take on a greater importance for the government. It requires to seek out the ways to align research agenda and its results with available strategic data on social and economic development of the Russian Federation and the ways to ensure its national security.

The ongoing transition to a new technological order, emerging drastic changes in energy technologies (e.g., new energy sources, the decentralization of energy supply through distributed generation, and smart energy systems), the anticipated transformation of world energy markets, and the ever-increasing integration of Russia into the global economy call for the consideration of most likely system-level consequences that these changes bring about to the national economy, as well as their effects on the energy and national security. This raises the question of extending the projection time frame to 2050 [2, 3].

Obviously, when dealing with such extended time frames, attempts at calculating numeric values and striving to boost their precision for most likely production and consumption volumes of specific fuel types become ultimately devoid of any practical meaning. At the same time, more consideration has to be given to the process of identifying emerging trends of the mutually conditioned development

Table 1 Stages and key objectives of the energy industry development prospects analysis

Stages	Objectives	Projections time frame	Adjustment interval
Long-term projection of the national energy system development	New trends, external conditions, and requirements for long-term development. Most likely challenges and difficulties. Narrowing down the uncertainty range. Documents that back up the energy strategy and strategic planning of social and economic development of the Russian Federation. Strategic vision for the required development of scientific knowledge, technologies, and pioneering research.	At least 25–30 years	On the ongoing basis
National energy strategy	Strategic vision of the efficient development required of the energy sector industries, including the priorities set for investment, technology use, economic, and innovation policies. Threshold values of energy security indicators and national security indicators. Principles of managing the energy sector development, a pricing policy, and a tax policy.	Up to 20–25 years	5–6 years
General schemes (master plans) and upgrading, and development programs for energy systems of individual industries	Working out reliable and efficient development options for energy industries. The options are subjected to given priorities within given constraints. Identifying prerequisites for such development to take place. Populating the pool of business proposals catered to prospective investors. Purchase orders for major equipment and engineering services as well as research assignments.	12–15 years	5–6 years
Regional energy programs	A well-justified fuel and energy balance. Development prospects of local energy and fuel supply sources and the energy infrastructure. Energy conservation plans and energy security enhancement programs.	12–15 years	5–6 years

(continued)

Table 1 (continued)

Stages	Objectives	Projections time frame	Adjustment interval
Long-term projections of the state of regional energy markets	Fuel and electricity price and demand behavior.	20–25 years	3–5 years
Strategic development plans by energy companies	Upgrading programs and capacity planning. Investment, pricing, technology use, and natural resource use policies.	Up to 10–15 years	3–5 years

of the energy industry and the economy at large, alongside most likely challenges and prerequisites for coping with them in a timely manner.

The objective and significant growth of uncertainty in both external and internal conditions of the energy sector development contributes to the increased importance of long-term projections but hinders the improvement of their performance. Hence, the necessity of developing further the methodology and projections. It is also obvious that such development should be underpinned by the system approach, while building upon the extensive hands-on experience of developing long-term projections as accumulated in the published research in Russia and abroad, as much as it should be informed by emerging trends of the mutually conditioned development of the national energy and economic systems.

The book undertakes an attempt at making long-term projections of the energy sector more well-grounded and practically useful by means of the following: defining the projection range (under various scenarios for exogenous factors) and dedicated methods for its analysis and incremental narrowing down of the uncertainty range. To this end, it is crucial to discern and solve the problems that are of the utmost importance for each of the projection time frame segments. The book provides an in-depth coverage of how to estimate and account for the correlation between the input data and the results it generates (the reliability thereof), on the one hand, and the projection time frame, on the other hand. The same applies to the issue of how to define what constitutes the acceptable forecast error. These matters are treated in Chap. 1.

The evolution undergone by the research tools that are employed in making projections as mapped against applicable time frames is covered in Chap. 2. That same chapter introduces an incremental approach to long-term projections of the national energy sector development as well as the approaches to analyze the entire projection range. Iterative calculations that make use of different models specific to each stage presuppose solving a set of key problems. The statement of these problems alongside the approaches to arrive at their approximate solutions is presented in the final chapters of the book.

The problem of the long-term price and demand projections for regional energy markets as well as the calculation of the price elasticity of demand as dependent on the nature of uncertainty with regard to future conditions are treated in Chap. 3.

The problem of estimating and accounting for the barriers that limit the space of possibilities for the energy sector development is dealt with in Chap. 4, while Chap. 5 that concludes the book is dedicated to the problem of the quantitative assessment of strategic threats and energy security indicators.

Irkutsk, Russia   Yuri D. Kononov

References

1. Federal Law of the Russian Federation dated June 28, 2014 N 172-FZ On strategic planning in the Russian Federation [Electronic Publication]. Retrieved from: http://www.rg.ru/2014/07/03/strategia-dok.html (In Russian)
2. Saenko VV, Kurichev NK (2013) Six steps of energy strategizing (the case of the energy strategy 2035/2050). Energeticheskaya politika 2:35–46 (In Russian)
3. Bushuev VV, Gromov AI (2013) Energy strategy—2050: its methodology, challenges, and opportunities. Energeticheskaya politika 2:11–18 (In Russian)

Contents

Chapter 1
The Effect of the Projection Time Frame on Projection Performance and Projection Performance Requirements

1.1 The Dependency of Input Data/Required Data on the Projection Time Frame

As the projection time frame extends into the future, it increases the uncertainty of future conditions of the energy industry development and compromises the performance of long-term projections, i.c. their reliability, precision, and value. The property of reliability is indicative of the probability that a projected variable will fall within the range predicted by the projection. The projection precision is defined as the width of the range of possible values of projected variables of the system. The value of a projection depends on how well-grounded the decisions are. Such decisions are to be made based on the numeric values assumed for projection variables [1, 2].

An overview of the dynamics of the projection reliability and precision as the projection time frame extends into the future is presented below as part of an analysis of the US energy system development projections (cases) published by the U.S. Energy Information Administration over the span of years from 1995 to 2013 [3, 4].

The discrepancy between projected values of the U.S. energy consumption indicators and the actual data for the year 2010, as shown in Fig. 1.1, provides evidence supportive of the assumption that even for the fuel and energy consumption projections that go as little as five years into the future, the projection error can reach 10% and more. That said, the assumption that the discrepancy decreases as the projection year nears the reference year does not hold true in all cases. However, the general tendency toward narrowing the projection error range appears well-established.

The deviations from the otherwise linear dependence of the projection error from the projection lead time could be explained away not only by the imperfection of projection methods but also by the idiosyncrasies of the reference year of 2010, a high volatility of the global oil prices in the given period, and the unstable economic growth.

Sections 1.1 and 1.2 are co-authored with Svetlana V. Steklova.

© Springer Nature Switzerland AG 2020
Y. D. Kononov, *Long-term Modeled Projections of the Energy Sector*,
Springer Geophysics, https://doi.org/10.1007/978-3-030-30533-8_1

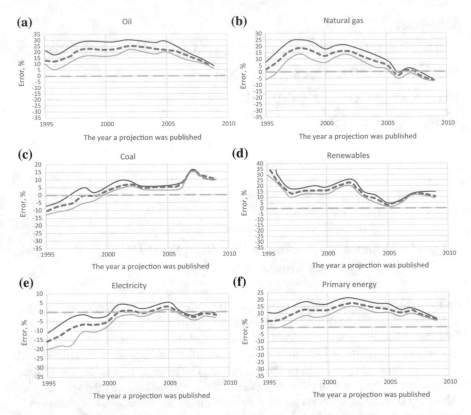

Fig. 1.1 a–f Discrepancy between projected and actual (as of 2010) values for the U.S. energy consumption by source (the dashed line within the range stands for the reference case values). *Source* Calculations by the author based on [6]

A key performance indicator of the projections, that at a given moment still extend into the future, is their reliability. The latter can be captured as the uncertainty range, i.e. the percentage difference between the highest and lowest values of a projected variable, as well as the coefficient of variation and dispersion (spread) of the values of projected variables set against their mean values or reference case values.

Published projections of the energy industry development in the USA and Europe that extend to 2030–2035 prove the non-linear nature of the escalation of the uncertainty range as the projection time frame extends into the future (see Figs. 1.2 and 1.3). The uncertainty range for primary energy consumption volumes in the USA for all cases and scenarios covered by the projections grows from the low 5–10% for the five-year time frame to the high 13–23% and 22–38% for projections that go out 15 and 25 years into the future respectively. The world energy market is undergoing a similar trend, but with a notably greater spread.

In "The Energy Strategy of Russia to 2030" as approved in 2009 (ES-2030), the difference between the total energy consumption under the worst and best case

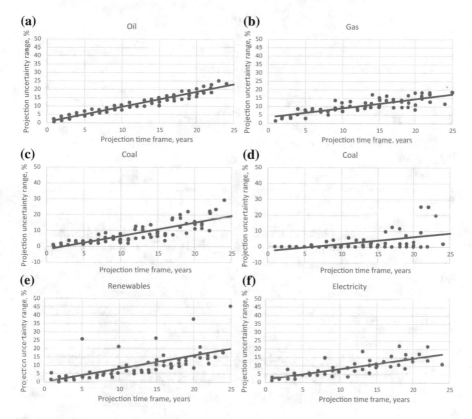

Fig. 1.2 a–f Correlation between the uncertainty range of projected energy consumption volumes in the USA and the projection time frame. *Source* Calculations by the author based on [6]

scenarios amounts to 7% for the first five years and subsequently grows to 22% and 31% for projections that go out 15 and 20 years into the future respectively.

The correlation between projection values of energy production and consumption in Russia, on the one hand, and the projection time frame t can be approximated by the following equations [5]:

- $I = 1.13\,t - 3$ ($R^2 = 0.74$), for primary energy consumption
- $I = 1.6\,t - 5.4$ ($R^2 = 0.89$), for electricity consumption
- $I = 0.75\,t + 1.5$ ($R^2 = 0.9$), for primary energy production

The above equations were derived from the data provided in ES-2020, ES-2030, and The General Scheme (Master Plan) for the Installation of Electricity Industry Facilities until the year 2020, with the projection time frame extended to the year 2030.

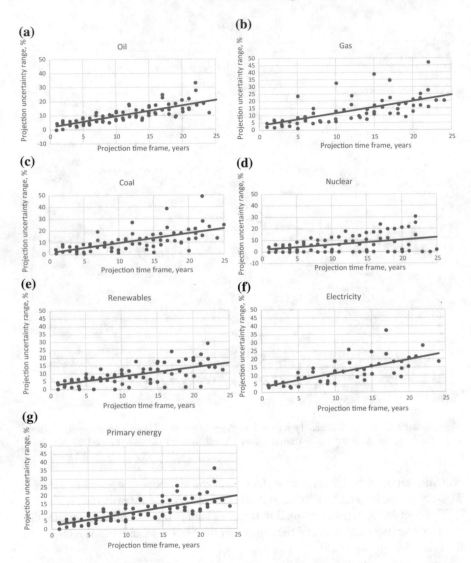

Fig. 1.3 a–g Correlation between the uncertainty range of projected energy consumption volumes in Europe (OECD) and the projection time frame. *Source* Calculations by the author based on [7]

1.2 Estimating the Acceptable Projection Error

The requirements that are set out for the projection accuracy as well as for the acceptable uncertainty range of the projected variable depend on how critical a given problem is and on the timing of the decision to be made based on the projection data.

When making investment decisions in the energy sector, one of the key problems is the valuation and risk assessment of large-scale capacity expansion projects in the electric power industry and the fuel industry. Such valuation has to be based on projections of the most-likely price and demand behavior on target energy markets.

The investor values the data on the more remote rewards and losses less than those of more immediate ones. This well-established fact is reflected in the ubiquitous use of discount rates embedded in the project valuation metrics: e.g. the Net Present Value (NPV), the Internal Rate of Return, the payback period, etc.

By varying the values of key input data variables for any of the time periods within the projection time frame and by tracing the effect of such variations on the project value, it is possible to arrive at conclusions bearing on the acceptable decrease in the projection accuracy as these time periods gradually go further into the future.

This approach was applied, in particular, to the sensitivity analysis of investment projects returns (NPV). First, we studied how the value of investment projects of the nuclear power plants construction responds to the changes that demand (production) volumes undergo over time. Second, we studied how the value of investment projects of the combined cycle gas turbines (CCGT) construction responds to the changes in gas prices. The simulation model employed for these calculations is capable of accounting for a comprehensive range of probability distributions of the input data within a given range of its possible values.

The calculation results (see Fig. 1.4) suggest a notable non-linear decrease in sensitivity of the nuclear power plant project value to the changes in electricity generation as the time periods go further into the future. Under the assumed input data, the surge in the demand for electricity by as high as 20% exerts an appreciable effect on the NPV only within the time frame limited by the first 15 years. Accordingly, projection performance requirements are relaxed for electricity demand projections at the end of the nuclear power plant life cycle.

Fuel price projection performance requirements get notably more relaxed as the time frame goes further into the future, which follows from the barely perceptible

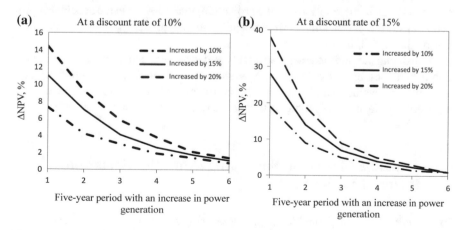

Fig. 1.4 a–b Correlation between the change in the nuclear power plant project's net worth and the increase in power generation in one of the five-year periods

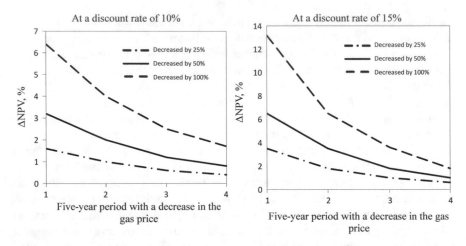

Fig. 1.5 Correlation between the change in the CCGT project's net worth and the gas price decrease in one of the five-year periods

effect of the gas price increase on the CCGT project value at the end of the projection time frame even in the case of the 1.5–2-time price increase (the NPV decrease then does not exceed 2–3%) (see Fig. 1.5). As is the case with the nuclear power plant construction project, the effect of the increase in the discounting rate on the net present value is perceptible only in early periods and is quickly attenuated afterwards.

Estimates of the likely error for the projection variables, when the latter are judged against the requirements set out with regard to their accuracy in making investment and other decisions, facilitate making more well-grounded the choice of the acceptable level of complexity of employed economic and mathematical models. Should an elaboration of these models, when compared against the currently employed models, lead to a change in the solution (i.e. the projection) that is less than the acceptable projection error, the practicality of implementing such approaches, however elaborate they might be, will be anything but obvious.

The existing methods to assess the quality and value of the input data, as well as the requirements set for such data by energy systems projections for various time frames, need further development. To this end, it is necessary to give due consideration to the features unique to the problems that are to be solved within a given projection.

1.3   Available Means for Narrowing Down the Uncertainty Range of Projections

The uncertainty of the future development of energy systems is conditioned by the uncertainty of their external environment and its fundamental properties: that is, complexity, pro-activity, agility, and non-linear interactions with the environment.

For long-term projections the key sources of uncertainty are the following ones: the magnitude of change in the raw material resource base (including extractable energy reserves), focus shifts in the scientific and technological advances, changes in the target orientation and makeup of the national economic system development, game-changing trends in the global economy and the export policy. As the projection time frame extends into the future, key uncertainty sources and factors shift from those technical and technological in nature towards social, economic, and institutional ones [6].

The fundamental possibility of narrowing down the uncertainty is implicated by the following: objective patterns and trends; subordination of the energy sector development to the requirements defined by the economic system; mutually conditioned operation of the systems that make up the energy sector; adjustments of macroeconomic performance indicators caused by the development of such systems and changes in energy costs (that is, addressing the issue of the negative feedback links); and, finally, by the system inertia inherent into the energy sector, which essentially holds back its drastic structural changes. That said, its successful implementation remains a challenging theoretical problem. An array of theoretical and practical approaches has been employed in Russia and abroad to cope with it. In Russia, the 1970s and 1980s saw a significant progress in capturing the uncertainty factor in projections of energy systems (see e.g. [7–10]).

In his recap of the development of long-term projections methodology during that period, academician Makarov [11] highlighted the uncertainty issue while noting that for the then available state-of-the art research methods it was only the "conc" of uncertainty of feasible energy systems development strategies (with the latter being subject to rapid divergence of boundaries as the time frame extended further into the future) that lent themselves to objective knowledge, which was moreover limited to key parameters only, i.e. total volumes and the overall make-up of energy production and consumption. He also advocated for introducing well-defined targets into the long-term projection that enables one to reduce the fuzzy "cone" with overestimated bounds of theoretically well-grounded trajectories to a more limited "bundle" of practically useful trajectories, i.e. a distinct strategy of energy systems development. To this end, the study of the "cone" of possibilities inherent in the energy industry development is subdivided into three stages [11]: (1) to identify so-called "boundary" strategies that limit what is possible or at least likely to happen within a given time period; (2) to lay bare the inner structure of the "cone" of inherent possibilities, i.e. clusters of essentially similar strategies that meet certain conditions of the energy industry development; (3) to unfold development trajectories for specific areas of the energy industry within each of the clusters. In doing so, the following problems are to be solved:

- to decide on the set of key factors that define the long-term development of the energy industry within the boundaries of the "cone" of intrinsic possibilities;
- to provide a quantitative assessment of the effects of these factors on the scale and structure of energy consumption and production;

- to interval-estimate possible changes in key variables of the prospective energy industry development that may be conditioned by the above factors;
- to sum up the findings of the above factor-wise analysis as a more general procedure for developing and estimating the parameters of the boundary strategies of the energy industry development.

An implementation of the multi-factor analysis of such complexity for the purposes of outlining the energy system prospects called for the development of dedicated models and methods that make up an interactive simulation system [12].

One of the most essential and widely used approaches to narrowing down the uncertainty of future conditions is to tether the energy sector development projections to scenarios of the national economy development and the state of the world energy markets.

Scenarios are a tool for making sense of the future, not of predicting it. They enable to arrive at a synopsis of possible development based on a logical and coherent system of assumptions about key driving forces and links.

The common practice of the projection analysis worldwide is undergoing a change toward increasing the number of scenarios covered by such an analysis. For example, the first projections published annually since 1979 by the U.S. Department of Energy covered only three cases for the 15-year time frame, while nowadays the number of cases has been increased to 30 scenarios that go up to 25 years into the future. This has grown into the dominating strategy to capture uncertainty and assess the effects of an array of factors on the U.S. energy industry development.

It is obvious that projections of the Russian energy sector development will benefit as well from increasing the number of the scenarios they cover, hence becoming more well-grounded (as of now, in practice the number does not exceed 3–4 cases). To this end, it is essential for the scenarios of the external conditions evolution not to be arbitrarily defined by experts but be based on in-depth projection studies. First of all, it concerns the projections of long-term social and economic development of the Russian Federation published by the Ministry of Economic Development of the Russian Federation as well as the projections of the world energy industry development and long-term international energy market trends for Russian exports.

To account for the uncertainty factor, it is common to have contingency calculations that measure the sensitivity of the energy sector structure to changes in input data for each of the scenarios. It should be pointed out that relying on such calculations of usually deterministic optimization models makes it difficult to address mutual dependency relations (correlations) between, say, such determinants as price and demand for energy carriers. Another shortcoming of deterministic models is that they are lacking in terms of accounting for the nature of the input data uncertainty. One of the most comprehensive and flexible methods available for the analysis of problems with potentially mutually-dependent random variables is the stochastic analysis based on Monte Carlo techniques [13]. It enables to capture any imaginable type of probability distributions for interval values of key input variables as well as the correlation between them.

The input data used for the projection analysis comes either as probabilistic (with known characteristics of probability distributions of input variables) or uncertain (probabilistic description is lacking). The former case sometimes calls for labor-intensive methods of stochastic programming to solve optimization problems. In the latter case, it is Monte Carlo experiments that are most useful when backed up by expert judgements bearing on probabilities of some of the events.

Other means that facilitate narrowing down the uncertainty and making the energy sector projections more evidence-based include: modeling the energy sector as a hierarchically organized system; improving the methods of modeling and capturing interactions between the energy industry and the economy at various hierarchical levels; clarifying requirements, conditions, and constraints relevant to each level based on the output of iterative calculations.

Further development and application of specialized mathematical methods and approaches also play an important part in "combating uncertainty". Such methods include subjective probabilities, sometimes applied in conjunction with utility functions [14]; the decision matrix and its subsequent developments [15]; algorithmic solving of optimization problems based on pre-defined decision criteria [14]; fuzzy sets and variables [16], and the like.

The evolution undergone by mathematical methods of capturing uncertainty is covered in the next section.

1.4 The Evolution Undergone by Uncertainty Handling Formalisms as Employed in Making Projections[1]

For the purpose of this overview it should be appropriate to highlight the following key building blocks that can be used to construct a wide array of approaches to combat the uncertainty: (1) an uncertainty handling formalism; (2) an aggregation technique; (3) a rational choice protocol under uncertainty (in projections, this item serves for modeling of the plausible behavior of economic agents). Based on these building blocks, one can construct arbitrarily complex procedures, integrate them into optimization models, and so on.

With respect to the input data uncertainty handling formalisms, one of the most salient tendencies is to radically revise the very notion of probability and of the uncertainty measure in general. The most seminal, theoretically elaborate, and widely applied attempts at the revision of probabilistic and post-probabilistic methods of this kind include the following ones: the Dempster-Shafer theory of evidence [17], the theory of imprecise probabilities [18], the possibility theory [19], the rough sets theory [20], as well as the application of interval analysis techniques, oftentimes used in conjunction with other formalisms to generalize them.

It is essential to point out that all significant differences notwithstanding, a defining feature that the state-of-the-art formalisms have in common is *the interval nature of*

---

[1]This section is authored by Svetlana V. Steklova.

estimates for the uncertainty measure championed by each of them. That said, the ranges of values defined by such measures have a *heterogeneous* structure to them with a clearly defined semantics that both of their boundaries are charged with, which makes them markedly different from straightforward interval estimates of some variable that one arrives at by way of expert judgements or otherwise. The particulars of the above semantics can vary widely depending on a given theory, however all of them share the concept of the lower bound as a guaranteed or justified value (as based on the available data and the accepted procedure of construing them), while the upper bound stands for a plausible value (the one that we have no reasons to exclude from our consideration): that is to say that the boundaries are inferred from the data themselves and are not exogenous to the model. The above duality of estimates follows directly from relaxing the Kolmogorov axioms and doing away with the additivity requirement of the measure (hence such uncertainty measures are labelled as *non-additive*): a hypothesis and its negation become to a certain extent independent of each other. Below we outline the key defining characteristics of each of the well-established uncertainty handling formalisms.

The Dempster-Shafer theory of evidence (also commonly referred to as "the theory of belief functions") construes the probability in an epistemology-minded way as a degree to which a hypothesis is proved based on the arguments that either support or refute it ("pieces of evidence"), each of which is allowed to be both unreliable and imprecise (the latter refers to the property of the argument to support not only a given hypothesis but also a set thereof without discerning which one of them is actually true). The analysis of the complete set of all such pieces of evidence available generates within the set of all hypotheses the so called "belief functions" (they stand for the degree to which a hypothesis is supported by available evidence) and their duals called plausibility functions (they stand for the degree to which a hypothesis is not refuted), and hence the values of these two functions form the above uncertainty interval. Another fundamental feature of the evidence theory is that it enables the rational synthesis of contradictory and unreliable information (as encoded by the above functions) by means of the Dempster rule of combination. Thus, the formalism accounts for new information not by just replacing the old values with the newer ones, but rather assuming their ambiguity (that of both old and new values) to "update" the beliefs we entertain in a logically coherent way (preserving in its "memory" the complete structure of our knowledge and its evolution over time), which is highly relevant when working with multiple sources of projections based on oftentimes mutually contradictory assumptions or even the assumptions unknown to us. As of time of writing, the theory is available in a whole range of its interpretations one of the most seminal and elaborated of which is arguably "a mathematical theory of hints" [21].

The imprecise probability theory is an evolutionary development of the Bayesian probability (that is to say that the subjective probability theory gets generalized so that its scope could encompass interval values as well) and just like the evidence theory it operates dual measures referred here as lower and upper "previsions" that are rooted in the behavioral interpretation of market games (as its metaphoric model, as is the case in the subjective probability theory, the imprecise probability theory draws upon the rational betting-game) as opposed to the epistemological stance of the evidence theory. The foundation of the theory was laid by Walley [22], in Russia the ideas germane to those behind the imprecise probability theory were developed by Kuznetsov [23]. The machinery of the theory is highly refined and, given its pedigree as that of a descendant of standard probabilistic methods, the theory is recommended for application in those areas that have traditionally employed probabilistic point estimates.

Alongside the multifarious developments in probabilistic methods and the extension of the scope of the very concept of probability, the recent years saw the emergence of theories that severe their formal relations with the concept of probability and introduce the measures that serve as alternatives to probability together with procedures to operate them. Among the most radical and seminal of all the ideas of this kind are the possibility theory (the name oftentimes used interchangeably with the fuzzy sets theory it is derived from) as well as the rough sets theory: these two developments, somewhat similar names notwithstanding, present a formal description of markedly different aspects of uncertainty.

The possibility theory operates the fuzzy set theory apparatus as its formal language and represents one of the pillars of the so called "soft computing" paradigm (on a par with artificial neural networks, genetic algorithms, and the like) [24]. The approach is catered first of all to the formal representation of the ambiguity of belonging of a certain element to a given class of elements and deals with the linguistic uncertainty inherent in any language used to describe real-life phenomena. At their most radical, fuzzy set techniques aim at making possible the computational operations with the words of natural languages [25]. In line with other approaches covered in this section, the possibility theory operates dual measures of uncertainty: the necessity measure (the lower bound) and the possibility proper (the upper bound). The fuzzy-theoretic apparatus has gained much popularity in applied studies and, in particular, was employed to generalize an approach based on the payoff-matrix [26].

The rough set theory is optimized for grasping the indiscernibility aspect of information uncertainty and is based on the premise that *the decrease in the accuracy of data makes them lend more easily to the identification of structures and patterns*, whereas the knowledge of "the world" can be reduced to the ability to distinguish between its entities and to classify them. To this end, the theory introduces the concepts of lower and upper approximations and the boundary region. As their output the models that employ the rough sets apparatus generate decision rules of the "if—then" type that are based solely on "raw data" (that is only on the information inherent in the input data themselves) without making any additional assumptions and imposing any structures for the data to fit in, unlike with both probabilistic techniques and fuzzy inference. The relevance and potential of using the rough set theory in the research related to the assessment of energy security indicators and their monitoring is obvious. It should also be pointed out that it was the rough sets theory that served as a backbone for the development of one of the most promising modern methods of multi-criteria optimization decision-making [27].

The presence of a range of values as an inalienable part of the uncertainty handling formalisms of today also contributed to the broadening of the area of application of interval analysis methods, most commonly, in hybrid models where the interval analysis is used to generalize other approaches including those key ones outlined above. It should also be noted that hybrid strategies are at the core of the most recent developments in the field. Widespread acceptance has been gained by all kinds of symbiotic constructs like fuzzy rough sets, fuzzy belief functions, interval-valued fuzzy sets, "gray" systems [28], and so on.

The development of uncertainty handling formalisms has triggered the corresponding development of the ways to aggregate uncertain information, that is to reduce its complex structure to a certain representative value, akin to the expected value, that can be employed to rank uncertain measures (that is, to compare intervals, belief functions, fuzzy sets, etc.), which is required if one is to make any decisions based on uncertain information. Among the most seminal efforts are arguably those that originate in the fuzzy set literature (presumably due to the pressing urge to develop new approaches to aggregation and "defuzzification" given that the expected value makes no sense whatsoever in the fuzzy realm), and yet are applicable to other uncertainty handling formalisms, which is a telling case of the "cross-pollination" that exists between the modern theories in the field. A key achievement is the OWA (ordered weighted averaging) operators theory [29] that serves as a generalization of an extensive class of aggregation functions and takes their form under certain values of predefined parameters with the latter being highly structured which makes it possible to "transition" between various classes of solutions. In particular, one can consistently control the attitudinal character of the operator, its "orness" (on the continuum between logical "AND" and logical "OR"), and its other properties.

The modeling of the rational choice as such, and hence the risk aversion of the decision-making agents, as "encoded" in the utility function, has long been formalized by means of the expected utility theory axiomatized in its modern state by Savage. However, the discovery of the paradoxes it entails, on the one hand, and the development of alternative uncertainty handling formalisms, on the other hand, proved the model as having severe shortcomings as both a descriptive model and even a normative model [30]. Eventually, even though the research on utility functions within the still mainstream Savage axiomatics remains active (utility functions classes, the risk measures they induce, and their parametrization are subject of numerous studies), cross-disciplinary studies that involve Behavioral Economics yielded a whole range of alternative theories that overcome the above paradoxes, among the most important of which are the cumulative prospect theory [31] as well the Choquet expected utility theory [32] that manages to integrate the expected utility theory into the apparatus of non-standard probabilistic and post-probabilistic measures.

As a general conclusion, it should be stated that the erstwhile probabilistic purism or that of any other kind in the best practices of today (and theorizing as well) has been replaced by freestyle task-specific constructing of hybrid "pipelines" that are meant to address any realistically formulated forecasting problems by turn-key combinations of model building blocks based on the above formalisms and those that were left behind the scope of this concise overview.

References

1. Pivovarov SE (1984) The methodology for comprehensive forecasting of the development of an individual industry. Nauka, Leningrad, p 192 (In Russian)
2. Siforov VI (ed) (1990) Prognostics: terms and definitions. Nauka, Moscow, p 56 (In Russian)
3. U.S. Energy Information Administration. Annual energy outlook (1995–2013) [Electronic Publication]. Retrieved from: http://www.eia.gov/forecasts/aeo/
4. U.S. Energy Information Administration. International energy outlook (1995–2013) [Electronic Publication]. Retrieved from: http://www.eia.gov/forecasts/ieo/
5. Galperova EV, Mazurova OV (2013) A study of a dependence of the growth of uncertainty of projections of energy production and consumption on the projection time frame. Energeticheskaya politika 3:33–38 (In Russian)
6. Shibalkin OY, Maiminas EZ (1992) Problems and methods of developing scenarios of social and economic development. Nauka, Moscow, p 176 (In Russian)
7. Makarov AA, Melentiev LA (1973) Research and optimization models for energy facilities. Nauka, Novosibirsk, p 274 (In Russian)
8. The uncertainty factor in making optimal decision in large energy systems: In 3 Vols. Ed. by L. S. Belyaev, A. A. Makarov. – Irkutsk: The Siberian Energy Institute of the Siberian Branch of the Academy of Sciences of the Soviet Union, 1974. (In Russian)
9. Belyaev LS (1978) Solving complex optimization problems under uncertainty. Nauka, Novosibirsk, p 126 (In Russian)
10. Belyaev LS, Rudenko YN (eds) (1986) Theoretical foundations of the energy systems analysis. Nauka, Novosibirsk, p 331 (In Russian)
11. Makarov AA (1988) On some of the problems of the long-term energy forecasting. National and regional energy systems: Governance theory and its methods. Nauka, Novosibirsk, pp 43–98 (In Russian)
12. Makarov AA (2010) Methods and results of forecasting the development of Russia's energy industry. Izvestiia RAN. Energetika 4:26–40 (In Russian)
13. Ermakov SM (1975) Monte Carlo methods and related problems. Nauka, Moscow, p 472 (In Russian)
14. Raiffa H (1977) Decision analysis: an introduction into the problem of decision-making under uncertainty. Decision Analysis. Nauka, Moscow, p 418 (In Russian)
15. Belyaev LS, Saneev BG (2010) Handling of uncertain information in the energy systems analysis. In: Voropai NI (ed) Energy systems analysis: the Siberian Energy Institute/Energy Systems Institute schools of thought in hindsight. Novosibirsk, pp 42–50 (In Russian)
16. Bellman R, Zadeh L (1976) Decision-making in a fuzzy environment. Problems of decision-making analysis and procedures. Mir, Moscow, pp 173–215 (In Russian)
17. Yager RR, Liu L (2007) Classic works of the Dempster-Shafer theory of belief functions. Springer, Norwalk, Conn, p 806
18. Augustin T. et al (2014) Introduction to imprecise probabilities. Wiley, Hoboken, NJ, p 448
19. Pytyev Yu (2000) Possibility. Theoretical and applied aspects. Editorial URSS, pp 192 (In Russian)
20. Pawlak Z (2013) Rough sets: theoretical aspects of reasoning about data. Rough sets. Springer, Dordrecht, p 231
21. Kohlas J, Monney P-A (2013) A mathematical theory of hints: an approach to the Dempster-Shafer theory of evidence. Springer, Berlin; New York, p 422
22. Walley P (1991) Statistical reasoning with imprecise probabilities, monographs on statistics and applied probability, vol 42. Chapman and Hall, London, p 706
23. Kuznetsov V (1991) Interval statistical models. Radio i Sviaz, p 352 (In Russian)
24. Bouchon-Meunier B, Yager RR, Zadeh LA (1995) Fuzzy logic and soft computing. World Scientific Pub Co Inc, Singapore; River Edge, NJ, p 497
25. Zadeh LA (1996) Fuzzy logic = computing with words. Trans Fuz Sys 4(2):103–111
26. Podkovalnikov SV (2001) Fuzzy pay-off matrix for under uncertainty for justifying decisions in the energy industry. Izvestiia RAN. Seriia Energetika 4:164–173 (In Russian)

27. Greco S, Matarazzo B, Slowinski R (2001) Rough sets theory for multicriteria decision analysis. Eur J Oper Res 129(1):1–47
28. Liu S, Lin Y (2010) Grey systems: theory and applications. Springer, Berlin, p 379
29. Yager RR, Kacprzyk J (1997) The ordered weighted averaging operators: theory and applications. Springer, Boston, p 347
30. Shafer G (1986) Savage revisited. Stat Sci 1(4):463–485
31. Tversky A, Kahneman D (1992) Advances in prospect theory: cumulative representation of uncertainty. J Risk Uncertainty 5. Advances in prospect theory 4:297–323
32. Schmeidler D (1989) Subjective probability and expected utility without additivity. Econometrica 57(3):571

Chapter 2
Further Development of the Long-Term Projections Methodology for the Energy Sector

2.1 The Evolution Undergone by Applied Models of the Energy Industry Development

The last quarter of the XX century saw long-term projections of the social-and-economic area and the science-and-technological area alike gaining importance in their agenda-setting role among academics. To a significant extent, this was facilitated by acute economic and energy challenges that became urgent back in the 1970s, alongside the long overdue recognition of oncoming threats of the scarcity of available resources and environmental concerns. These global threats were demonstrated, in particular, in publications known as "Reports to the Club of Rome" [1] that were based on system dynamics models.

The due recognition of the important part played by long-term projections in developing strategic decisions was instrumental in the development of projections methodology based on the systems analysis methods. A major contribution to the development of such methods and economic and mathematical models as applied to the energy industry was made by the Siberian Energy Institute of the Siberian Branch of the Academy of Sciences of the USSR (nowadays, the Melentiev Energy Systems Institute of the Siberian Branch of the Russian Academy of Sciences) [2].

It should be pointed out that system studies commenced in the USSR back in the 1960s and 1970s by the end of the 20th century have joined the stock of mainstream tools worldwide, while a number of variations on these methods and models have been applied in many countries across the globe.

The evolution of the models and their applications initially followed the path of gradual sophistication by way of an increasingly more elaborate representation of the economy in energy models and a more detailed description of the energy sector in models of the economy. However, by the 1970s, in the USSR and abroad, the concept of employing hierarchically-built models to account for the interactions between the energy industry and the economy gained wide acceptance, with macroeconomic

© Springer Nature Switzerland AG 2020
Y. D. Kononov, *Long-term Modeled Projections of the Energy Sector*,
Springer Geophysics, https://doi.org/10.1007/978-3-030-30533-8_2

models coming into ever more prominence. Abroad, it was mostly econometric models built upon the notion of the general equilibrium, while in the USSR and other planned economies the input–output model served the same purpose.

Those same years saw the development of a system of economic and mathematical models for long-term projections of the energy industry development that was made up of the following: an optimization model of the national energy sector development [3], an input-output optimization model (MIDL) [4], a regression model of energy consumption, and a model of the requirements set by the energy sector with respect to its development that were to be met by linked industries and production facilities (IMPAKT) [5]. A structurally similar system of models was employed for the purpose of making the world energy projections at the International Institute for Applied Systems Analysis (IIASA), Laxenburg, Austria [6].

Nowadays, cross-national differences in modeling interactions between the energy industry and the economy have grown much less pronounced, while the methods addressing the issues specific to market economy workings have become more up to the task.

The changes brought about by the 1990s to the social environment and business administration set a task of revamping the time-honored methods and tools employed for the energy sector projections. The new economic order made it indispensable to account for emerging energy markets and their role. Market relations set the problem of decision-making criteria anew. Incorporating the procedures of the state regulation of the energy industry development became an important task [7].

As the scope of problems extended with the complexity of the tasks growing further, the tendency toward building computational systems that imply the use of capable computers and the state-of-the-art information technology was becoming more and more conspicuous. The two basic approaches to the problem have come to dominate the discourse.

The first one is based on the ad hoc choice of certain models out of a pre-defined pool populated by a variety of economic and mathematical models, so that the selected models would be deemed indispensable for solving specific problems of long-term projections of the energy industry development. In doing so, the interactions between models in the newly built systems do not have to be automated, while their joint use does not imply joint optimization. Such an approach has been implemented, for example, at the Melentiev Energy Systems Institute of the Siberian Branch of the Russian Academy of Sciences, where the experience accumulated over several decades helps to tweak the models in a sensible way and build their combinations tailored to specific projections. The adjustment of input data and constraints during the controlled iterative calculations and harmonizing various models is instrumental in solving the problem of accounting for and reconciling of multiple optimality criteria.

An alternative approach to model integration is to automate calculations and to use a unified database (the integrating module) and even a shared optimality criterion. This approach is epitomized by the NEMS (The National Energy Modeling System) system [8]. It was designed and implemented by the Energy Information Administration (EIA) of the U.S. Department of Energy (DOE) in 1993 and have

since then been used to project possible consequences of alternative cases of various (probable) energy market conditions for the energy, the economy, the environment, and the national security.

NEMS is made up of over dozen models (modules), including but not limited to the following: the macroeconomic activity module, the international energy module, four supply modules; two conversion modules; four end-use demand modules, etc. (see Fig. 2.1 for the high-level structure of the modeling system). To this end, the system ensures an equilibrium of supply and demand for energy carriers for 9 aggregated regions covering all U.S. states. Each industry-specific component module is self-contained with respect to calculating supply and demand for a given energy carrier. The information is then fed back to the integrating module that is elaborately structured.

Computer modeling systems employed to study energy planning and environmental management problems have been developed in the EU countries as well. One of them is the MESAP-III (Modular Energy Sector Analysis and Planning system), an information and decision support system for energy and environmental planning computer modeling system maintained by the Institute of Energy Economics and Rational Energy Use (Stuttgart University, Germany) [9] that is known to have found its application in a number of European and Asian countries. The system consists of six modules (models), an automated network database, three central databases that serve as information exchange hubs, and is meant to evaluate the demand for useful and end-use energy, optimization of the fuel and energy supply system, identification

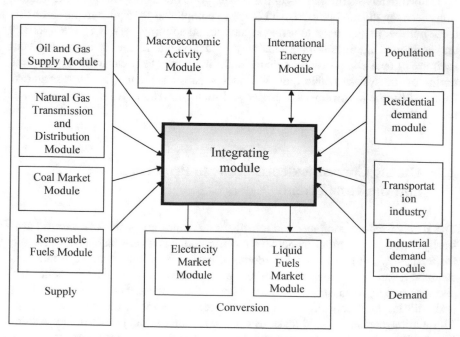

Fig. 2.1 The NEMS modeling system of the U.S. Department of Energy

of environmental concerns, determination of the amount to be invested and the return on investment, an analysis of direct and reverse energy-economy links.

In Russia, new modeling and information systems that are based on the integration of existing and newly built mathematical models underpinned by the cutting edge information technology are developed to ensure systemic valuation and implementation of risk assessment for various scenarios of the energy industry development, with the latter treated as an integral part of the economy. Such models are meant to enable us to capture possible consequences of decisions that are being worked out by top-level political and economic government agencies of the country.

One such system that proved successful is SCANER [10] that is being developed and maintained by the Energy Research Institute of the Russian Academy of Sciences (see Fig. 2.2).

Alongside vertical (cross-level) interactions, the SCANER modeling system accounts for strong horizontal links, i.e. those between regional energy sector models, fuel and energy sector industries and companies, as well as between functional (that is, product demand, production and transportation, economy and financing, etc.) and temporal modules of a model of the same economic entity. Dedicated methods of horizontal harmonization of solutions obtained by optimization models have been developed to account properly for such links. They provide for iterative exchange of information between a given model and all the other models of the same hierarchical level on production (consumption) volumes and prices (cost-effectiveness) of each energy source used in each of the regions in each of the years of a given time frame.

Unfortunately, as the authors of the system point out themselves [10], the interactions between the models of SCANER are so complex as to require an unreasonably large number of iterations to arrive at an optimal solution by way of their vertical and horizontal algorithmic coordination with an acceptable level of accuracy. Hence their use of heuristic procedures alongside formal methods to ensure first the overall tuning of the system within quite a limited area of solutions (that is, a scenario), followed by the arrangement of major and minor iterative cycles within the same area.

2.2 On the Correspondence Between Projections Methods and the Time Frame

The state-of-the-art computer and information technology enables us to build arbitrarily complex systems of models. That being said, it is unreasonable not to account for the following: large and ever growing uncertainty of input data; dependence of required accuracy of calculations on the time frame and the problem-specific context; complexity of construing the results under enormous number of variables, links, and criteria; the practicality of involving experts in some of the calculation steps. These idiosyncrasies should make us cautious when building multi-model systems for simultaneous (joint) optimization of energy and economy development.

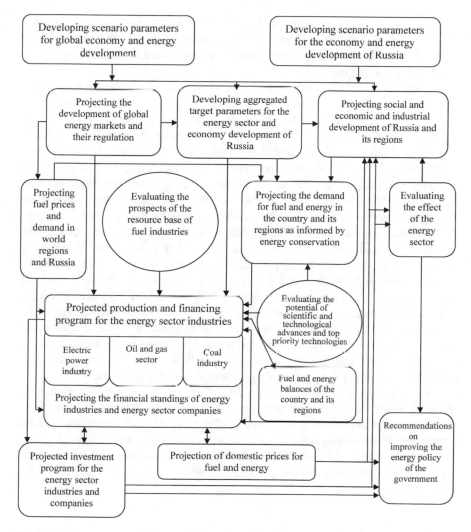

Fig. 2.2 Basic modules and information flows of the SCANER modeling and information system

Such systems that feature fully automated calculations are not only a pain to debug, but, what is of more importance, do not lend themselves naturally to tracking the contributions of individual variables and links and interpreting the results afterwards. An informal approach that implies an analysis of the information that serves as the output of one model and is then fed into another model significantly simplifies the research of complex problems.

An important principle of model improvement that academician Lev A. Melentiev held in high esteem [11] is matching the accuracy of calculation results with the accuracy of the input data. The principle is similar to the nearly proverbial Occam's razor principle and assumes building models that are as simple as possible yet capable of accounting for defining properties of the system being researched that are required to appropriately solve the task under given conditions. This echoes the following quote attributed to Albert Einstein as well: "Everything should be made as simple as possible, but not simpler" [12].

Striving for an utterly comprehensive while mathematically tractable treatment of development dynamics and non-linear relationships within the system studied as well as the detailed representation of its structure can go against the grain of the inherent uncertainty of input economic data and the mutable nature of properties of complex systems that are being modeled, which can even entail negative outcomes.[1]

The principle of correspondence between research tools and the actual uncertainty of the input data fed into them as well as the required degree of projection's accuracy has so far been implemented based on intuitions held by model developers and model users and remains more of an art than a science. A more well-grounded approach to the implementation of this principle can be arrived at by way of a quantitative analysis and a trade-off between the performance of projections and their value as an input into the decision-making process as dependent on a given time frame.

Strong dependence of the results of mathematical modeling on the input data quality and assumptions, a lack of accounting for political, regulatory, and other contingencies have sparked interest in foresight methods (under the increasing uncertainty of the future). Such qualitative and semi-qualitative methods rely on integrating the knowledge of the significant number of experts. Their contribution increases as the projection's time frame extends further into the future and they start to dominate other inputs for overly long-term projections of the energy industry.

An intermediate zone between mathematical modeling and foresight methods is occupied by the method of structural forecasting [14]. The latter implies building a logical model of key factors and conditions of the energy industry development and an analysis of logical contradictions that arise in the process of such development. This method is backed by results of orthodox input-output model forecasting and generalizes them while highlighting sustainable tendencies and invariant trajectories of future development.

2.3 A Novel Incremental Multi-stage Approach to Long-Term Energy Sector Projections

For the sake of convenience, it is reasonable to divide any energy sector projection into three major steps and several stages each, in turn, subdivided into the problems

---

[1] Academician Nikita N. Moiseev, while referring to international practices of building models for projections of the economic development, pointed out that creating ever more accurate models, striving for accounting for the increasingly large number of internal links and details of a given process yield negative results, i.e. the more accurate model representations were, the worse their predictive power was [13]. Arguably, a statement by a renowned mathematician René Frédéric Thom, known as the founder of catastrophe theory, relates to modeling economic systems of such complexity: "The more rigorous the less meaningful".

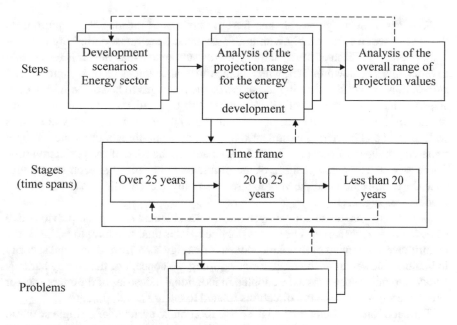

Fig. 2.3   The flowchart of the energy sector projections development

that are specific to it and are deemed most important within their scope. Solving these problems makes projections more defensible and practically relevant. The iterative alignment and adjustment of calculation results at each step and each stage also contribute to this end (see Fig. 2.3).

Step 1 is to come up with scenarios that reflect possible external conditions underpinning the energy sector development and the requirements that it has to meet with respect to economic, social, geopolitical, and environmental factors. Such scenarios consolidate various well-grounded opinions and projections of the development of the economy, the state of international energy markets with respect to Russia's exports, as well as the scientific-and-technological advances. The methods of developing the projections of this kind have a number of features unique to them. An analytical review of publications by relevant institutions and experts is instrumental in developing scenarios.

It is essential at Step 1 to project the demand for energy carriers (electricity, central heating, boiler and furnace fuels, motor fuel, as well as the fuel used as a raw material). Such a projection should be based on scenarios of economic development and established trends and patterns. A more granular treatment of the demand for energy carriers, its breakdown by fuel types, accounting for the price elasticity of demand and other factors all take place at Step 2, which is the main step for making a projection.

It is important to point out that all parameters of each scenario should be assigned as interval estimates rather than point estimates.

Step 2 is to identify possible and feasible options of meeting the requirements the energy sector is to comply with under assumed conditions, delimiting the projection range (for each of the scenarios) and its subsequent analysis. It is reasonable to break this step down further, which is the most labor-intensive of all, into several *stages*, thus gradually extending the scope of the problems to be solved within each stage and increasing the complexity of the iterative calculations procedure to accommodate the interactions between prices, demand, and energy production volumes, as well as financial barriers, investment risks, and strategic threats in an increasingly more comprehensive way (see Table 2.1). Changing the span of the projection time frame as well as the composition of the problems to be solved at each stage influence the composition and the very nature of employed economic and mathematical models (see Fig. 2.4).

Step 3 is to generalize and analyze the results obtained during the previous two steps for various scenarios. The results obtained at this final step have to facilitate the quantitative assessment of threshold values for energy security and national security indicators, and serves as the basis for developing the concept of the energy strategy and development programs to be adopted in individual industries of the energy sector and make a case for research directions related to energy development.

The set of options that make up the projection range even under a single scenario is determined not only by the ambiguity of the assigned parameter values, but also by the possibility to employ different criteria in calculations.

Vector optimization enables us to obtain and apply integral criteria based on the aggregation of various criteria. Formally speaking, such a criterion is a certain analytical expression that defines in quantitative terms its dependence on a system of local objectives and their types [15]. This way of arriving at the performance criterion may prove seminal for projections of the development of the national economy or the scientific-and-technological advances, especially in the case of accounting for the reallocation of weights within the system of local criteria over time, with a decrease in the importance of some of them as the projection time frame extends further into the future.

When it comes to the practical implementation, the inherent multi-criteria nature of economic systems gives way to choosing a single most important criterion, while the rest of them serve as boundaries of the allowable range of values of alternations in key factors. In optimization models of the energy sector, such criterion boils down to minimizing net present costs (inclusive of the investment component) that are required to meet a given demand for energy carriers.

In the case when projections of the energy sector development are made by means of a system (hierarchy) of models with different objective functions, the problem of accounting for the multi-criteria nature of the problem is to a significant extent solved through linking these models to iterative top-down and bottom-up calculations and harmonizing the features and goals of these systems that are specific to individual industries and regions.

Relying on the minimum cost criterion (that is, levelized costs) in optimization models of the energy sector and its industries is justified when the results of the economic performance assessment for the most important of the newly built facilities

Table 2.1 Features of individual stages of the energy sector projections under a predefined scenario of the economic development

Stages	Key problems	Key input data	Projection time frame	Employed models (model systems)
Initial stage	The projection range for potential development of the national energy sector. Delimiting invariants and the risk and uncertainty area. Preliminary assessment of fuel price dynamics and potential barriers to the development of individual energy industries.	Demand for energy carriers. Global energy markets prices and potential exports. Time constraints on the development of centers of fuel production and new energy sources and the profiles thereof. Technical and economic performance indicators of new technologies.	Over 25 years	Key indicators: Performance of the energy sector, the electric power industry, and regional energy markets
Final stage	Narrowing down the uncertainty area of the projection range. Projection of the state of regional markets. Quantitative assessment of strategic threats. Calculating threshold values for energy security indicators under a given scenario.	Data and results of the previous stages.	Up to 20 years (inclusive of a delayed effect)	Disaggregated basic models and supplementary models: models of the energy sector industries, national economy, rational exports, regional energy supply, investment behavior patterns of energy companies.

(as determined by the energy companies module) are accounted for e.g. by way of imposing constraints (see Fig. 2.4). However, it is still possible to determine the effect of the results of optimization of the state of the energy markets projections without including the latter module in the overall projection-making workflow. This can be achieved by using the maximum profit criterion (given the difference between market prices and levelized costs) for one of the iterations of the energy sector model.

An essential component of any projection is to outline a potential range of projection values (e.g. for new capacity additions dynamics, the share of individual

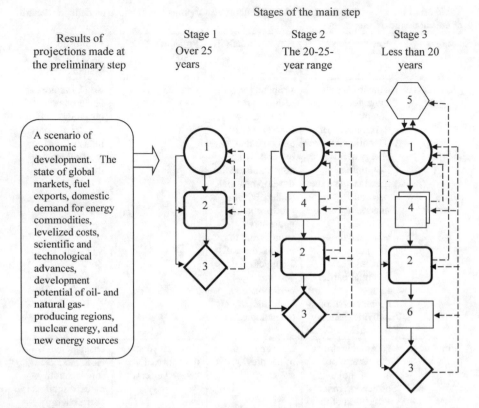

Fig. 2.4 The composition and interaction of models at different stages of developing and studying long-term development options available to the energy sector. Legend: 1—energy sector; 2—regional energy market trends (demand and prices); 3—barriers and threats; 4—energy sector industries; 5—macroeconomic conditions; 6—energy companies

industries in the energy sector makeup and the fuel and energy balance), to delimit invariant and unstable areas (risk zones), and to assess the magnitude and the time of exposure to strategic threats and major issues.

2.4 Structural Analysis of the Projection Range[2]

The purpose of the structural analysis of the projection range is to delineate its boundaries and rank all the facilities that fall within it by the likelihood of their making it to this range.

---

[2]The section is co-authored with Svetlana V. Steklova and Dmitry Yu. Kononov.

Invariants are defined as robust and most likely solutions that are arrived at as a result of contingency calculations under expected changes in input data and various potential scenarios of the system development. Essentially, invariants define the lower bound of the so named 'uncertainty cone', in the case when the latter is delineated by the very lowest values of a given variable across multiple cases. In the context of the energy sector projections, to delimit invariants is to search for robust structures in the fuel and energy balance and the energy sector industries, and to arrive at interconnected values of the very minimum demand for energy carriers or new capacity additions in the electric power industry and the fuel industry. Delimiting invariant solutions facilitates arriving at the so called "guaranteed" development strategies.

Optimal solutions and facilities that are included only in a subset of all cases (scenarios) define the instability area of the projection range. For example, the newly built power plants with the capacity that is higher than the bottom end of the uncertainty cone all belong to this area. The more this value differs from the invariant one, the higher the risk and the lower the likelihood is with respect to adding power plants of this type and capacity within a given time frame. The top end of the projection range is defined by the facilities that are most risky in this sense.

Obviously, the risk level has to be measured with respect to meeting the requirements (criteria, constraints) that were accounted for when solving the problem of selecting rational options under given conditions.

The choice of techniques of the projection range analysis and the quality of its results depend to a large extent on the number and defensibility of the energy sector development options, yet none of them can do away with expert judgements.

The part played by experts is especially important for the analysis and juxtaposition of available projections published by various agencies and authors that differ in their release dates, techniques, and input data assumptions. In such case, one should assign more weight to the more substantiated and up-to-date projections. In doing so, priority is given to the projection range (that is, its lower and upper bound values) of the growth rate and the rate of structural changes, rather than to the cardinal values of anticipated production for individual energy carriers or capacity additions.

This approach was applied to an analysis of the projection range of nuclear power plants prospects in Russia. We had at our disposal a total of 9 published long-term projections made between 2009 and 2014 by the Energy Research Institute of the Russian Academy of Sciences [16], the Melentiev Energy Systems Institute of the Siberian Branch of the Russian Academy of Sciences, the Agency for Forecasting Balances in Electricity Sector (JSC APBE) [17], and other agencies, as well as the General Scheme (Master Plan) for the Installation of Electricity Industry Facilities until the year 2020 [18] and the Energy Strategy of Russia to 2035 (hereinafter referred to as ES-2035) [19]. A wide disparity revealed across these projections can be explained away, first and foremost, by changing assumptions regarding the economic growth rate. To narrow down the projection range as demonstrated in Fig. 2.5 and in Table 2.2 we highlighted the projections extracted from ES-2035.

It follows from the analysis of the above sample data that, to give just one example, the minimum (that is, invariant) total capacity additions for nuclear power plants in

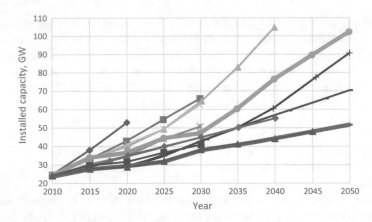

Fig. 2.5 The projection range for the nuclear power development prospects in Russia. *Note* The area encompassed by the two bolder lines is that covered by the ES-2035 projections [19]. *Source* Projections published in years 2009–2014 as compiled by the author

Table 2.2 Performance indicators for nuclear power plants prospects as per projections published in Russia in 2009–2014

Performance indicators	Unit	2010 (actual)	Overall projection range			ES-2035		
			2020	2030	2040	2020	2030	2040
Average annual growth rate	%	n/a	1.8–5.1	2.8–4.9	1.5–5.1	2.0	3.1–4.0	1.5–3.6
The share of total installed generation capacity	%	11.1	11–14	14–18	14–24	11.4–11.5	13.5–15	14.2–18.5

Source Compilation of published projections

the years 2015 to 2040 are equal to 20 GW (the average annual capacity additions amount to about 0.8 GW), while their share in the overall makeup of power generation capacity is projected to be at least 14% (cfr. 11% in 2011), and the minimum average annual growth rate over this period will reach about 2%.

To arrive at the upper boundary of the same indicators on the basis of available projections is more challenging: total capacity additions over the 25 year time period vary from the low 30 GW as per "the innovations-driven scenario" of the ES-2035 to the high 78 GW as per the maximum electricity consumption case developed by the Energy Research Institute of the Russian Academy of Sciences (published in 2009 [16]).

The share of nuclear power plants in the total power generation capacity by the year 2040, according to the above scenarios, may reach 18.5% and 24% respectively, under the average annual addition rate of 3.5% and 5% (Table 2.2).

It should be pointed out, that according to the world energy projections published in 2013 in the USA [20], the share of nuclear power plants in the overall structure of power plant generation capacity in Russia is closer to the values assumed in the ES-2035 innovations-driven scenario, growing from 13.3% in 2020 to 15.1–15.5% in 2030 to 16.5–18.2% by 2040. That said, the values projected by the U.S. agency for nuclear power plant capacity additions will exceed the values assumed in the ES-2035 by 2–6 GW by year 2030, while lagging behind the latter by 6 GW during the 2035–2040 time period.

It is obvious, that of all considered projections of the Russian energy sector development it is the scenarios assumed in the Energy Strategy that are of most value and are eventually most likely to realize.

Given sufficiently representative and reasonable set of cases (scenarios), the probable inclusion of any facility of a certain capacity X_i within the projection range can be estimated in a more well-grounded way based on the share of all cases that assume the capacity greater or equal X_i in the total number of cases.

One arrives at such cases, however, under conditions and scenarios that are of different likelihood to begin with. If it is deemed feasible to assign weights proportionate to such relative plausibility by way of expert judgement or otherwise (e.g. by using the 10-item Likert scale or the like) and normalize them between 0 and 1 afterwards, then the probability that a given facility would appear in the projection with its capacity equal to X is defined as per the equation below:

$$v(t) = n_i w_i / n,$$

where n—the total number of cases covered by the projection, n_i—the number of cases with projected capacity for a given facility is greater or equal than X, w_i—the sum of normalized weights of such cases. The higher the probability of occurrence X, the lower the risk of such an entity (case) not to appear in the projection range, hence:

$$R(X) = 1 - n_i w_i / n.$$

The above concept or risk assessment was tested against the problem of new capacity additions for combined cycle gas turbines (CCGT) in the USA. We studied the 30 cases of the U.S. energy sector development published in 2013 [21].

The projection range for new capacity additions of CCGTs from 2015 to 2040 is shown in Fig. 2.6. By as early as 2026 the difference between maximum and minimum projected capacity values exceeds 100 GW.

An overview of the cases (Table 2.3) proves their heterogeneous nature, unequal plausibility and significance. Therefore, we had to assign them different weights based on our expert judgements. To this end, we assigned the maximum 10 point weight to the reference case, while the minimum weigh (1 to 2 points) was assigned to the unlikely cases the only purpose of which was to estimate the impact of specific factors (e.g., maintaining the energy efficiency at the present level, etc.) on the electric power industry prospects.

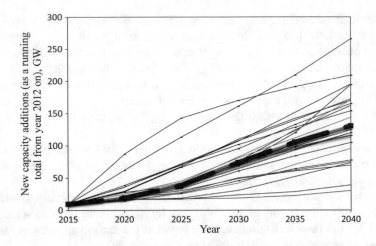

Fig. 2.6 Projected dynamics of new CCGT capacity additions in the USA under various scenarios of the energy industry development. *Note* The reference case is outlined in bold. *Source* Published data [22] as compiled by the author

Table 2.3 Scenarios of the U.S. energy industry development through year 2040

No.	Case	Summary	Weight
1	2	3	4
1	Reference case	Real GDP grows at an average annual rate of 2.4% from 2011 to 2040. Crude oil prices rise to about $141 per barrel in 2040.	10
2	High economic growth	Real GDP grows at an average annual rate of 2.8%. Other energy market assumptions are the same as in the Reference case.	9
3	Low economic growth	Real GDP grows at an average annual rate of 1.9%. Other energy market assumptions are the same as in the Reference case.	9
4	High oil price	High oil prices ($204 per barrel in 2040)	8
5	Low oil price	Low oil prices (fall to $70 per barrel in the short run and then recover to $75 by the end of the projected time frame).	8

(continued)

Table 2.3   (continued)

No.	Case	Summary	Weight
6	No sunset	Begins with the Reference case and assumes extension of all existing energy policies and legislation that contain sunset provisions.	2
7	Extended policies	Begins with the No Sunset case but excludes extension of the ethanol and biofuel subsidies that were included in the No Sunset case, as well as stricter energy conservation requirements.	2
8	Accelerated nuclear retirements	Assumes that all nuclear plants are limited to a 60-year life, while uprates are limited to 0.7 GW.	3
9	Electricity: high nuclear	Electrical power industry development with a higher share of nuclear power.	4
10	Accelerated coal retirements	Combines the High Coal Cost case with increasing operating costs for existing coal plants.	4
11	Accelerated nuclear and coal retirements	Combines the assumptions in the Accelerated Nuclear Retirements and Accelerated Coal Retirements cases.	3
12	Electricity: low nuclear	Electrical power industry development with a lower share of nuclear power.	4
13	ESICA legislation	Begins with the Reference case and assumes passage of the energy efficiency provisions in S. 1392, including appropriation of funds at the levels authorized in the bill.	2
14	Low coal cost	The coal industry development under the assumption of lower coal mining costs. Regional productivity growth rates for coal mining are approximately 2.3 percentage points per year higher than in the Reference case, and coal miner wages, mine equipment costs, and coal transportation rates are lower than in the Reference case, falling to about 25% below the Reference case in 2040.	4

(continued)

Table 2.3 (continued)

No.	Case	Summary	Weight
15	High coal cost	The coal industry development under the assumption of higher coal mining costs. Regional productivity growth rates for coal mining are approximately 2.3 percentage points per year lower than in the Reference case, and coal miner wages, mine equipment costs, and coal transportation rates are higher than in the Reference case, ranging between 24% and 31% above the Reference case in 2040.	4
16	Renewable fuels: Low renewable technology cost	Capital costs for new non-hydro renewable generating technologies are 20% lower than Reference case levels through 2040.	2
17	Low oil and gas resource	Estimated ultimate recovery per shale gas, tight gas, and tight oil well is 50% lower than in the Reference case.	3
18	High oil and gas resource	Estimated ultimate recovery per shale gas, tight gas, and tight oil well is 50% higher and well spacing is 50% lower than in the Reference case. Assumes the development of tight oil and other unconventional oil resources.	4
19	High vehicle miles traveled (VMT)	Assumes increases in VMT per licensed driver. VMT per licensed driver is 3% higher than in the Reference case in 2012, increases to 7% above the Reference case in 2027, and decreases back to 3% above the Reference case by 2040.	3
20	Low vehicle miles traveled (VMT)	The Low VMT case includes assumptions for a decline in licensed drivers, as well as decreases in VMT per licensed driver. VMT per licensed driver are 5% lower than in the Reference case for the entire projection.	2

(continued)

Table 2.3 (continued)

No.	Case	Summary	Weight
21	High rail liquefied natural gas	Assumes a higher LNG locomotive penetration rate into motive stock such that 100% of locomotives are LNG capable by 2037.	2
22	Low rail liquefied natural gas	Assumes a lower LNG locomotive penetration rate into motive stock, at a 1.0 average annual turnover rate for dual-fuel engines that can use up to 80% LNG.	2
23	2013 demand technology	Assumes that future equipment purchases in the residential and commercial sectors are based only on the range of equipment available in 2013. Commercial and existing residential building shell efficiency is held constant at 2013 levels.	1
24	High demand technology	Assumes earlier availability, lower costs, and higher efficiencies for more advanced residential and commercial equipment. Assumes significant efficiency improvements over the Reference case for all sectors.	4
25	Best available demand technology	Assumes that all future equipment purchases in the residential and commercial sectors are made from a menu of technologies that includes only the most efficient models available in a particular year, regardless of cost. New and existing commercial building shell efficiencies improve 50% more than in the Reference case by 2040.	2
26	No greenhouse gas concern	No GHG emissions reduction policy is enacted, and market investment decisions are not altered in anticipation of such a policy.	2
27	Greenhouse gas $10	Applies a price for CO_2 emissions throughout the economy, starting at $10/metric ton in 2015 and rising by 5%/year through 2040.	2

(continued)

Table 2.3   (continued)

No.	Case	Summary	Weight
28	Greenhouse gas $25	Applies a price for CO_2 emissions throughout the economy, starting at $25/metric ton in 2015 and rising by 5%/year through 2040.	2
29	Greenhouse gas $10 and low gas prices	Combines GHG10 and High Oil and Gas Resource cases.	2
30	Low electricity demand	Begins with the Best Available Demand Technology case, which lowers demand in the building sectors, and also assumes greater improvement in industrial motor efficiency.	2

Note The summaries of cases come from [19], their weights are estimates by the author

Table 2.4 shows the risk values of scenarios (the most important ones) of new CCGT capacity additions calculated as per the above risk measure. Figure 2.7 shows corresponding risk distribution as a function of CCGT capacity additions for each snapshot in time. The dependency lends itself reasonably well to logarithmic approximation (the coefficient of determination takes the values close to 0.8).

Table 2.4 Risk values of projected CCGT capacity additions in the USA (in select cases), %

Case	Year				
	2020	2025	2030	2035	2040
1	50	33	57	58	53
2	81	81	89	86	76
3	61	10	10	16	16
4	14	59	30	32	26
5	14	59	30	32	26
9	31	26	26	26	41
10	89	94	86	83	87
12	77	30	46	70	91
13	34	47	44	51	63
14	3	52	50	46	46
15	69	77	73	67	67
17	0	18	18	14	14
18	73	73	77	74	84
24	6	6	6	6	8

Note Estimates are made based on the above method of risk assessment for given CCGT capacity additions applied to the cases with the weight greater or equal than 4 points

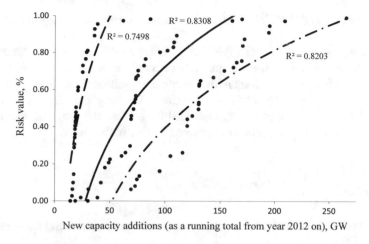

Fig. 2.7 Risk values for CCGTs of varying capacity, USA. *Source* Based on the analysis of all 30 cases covered in [20] as performed by the author

Table 2.5 outlines the overall projection range (based on 30 cases) for the generation capacity composition, alongside the projection range based on the less risky cases. It shows that the CCGT share in this make-up while being second to coal-fired power plants in 2012 will eventually start to dominate it in the years to follow. Under all cases it is projected to be at least 22–23%, with the peak value reached by year

Table 2.5 Dynamics of the generation capacity makeup in the U.S. Department of Energy projections, %

Power plant type	Actual	Overall projection range (all cases)			Most likely projection range (14 cases)		
	2012	2020	2030	2040	2020	2030	2040
Coal-fired power plants	29.7	10.8–26	4.6–24.5	1.5–24	21.9–25.9	18.8–24.5	16.5–24
Gas-and-oil-burning power plants (steam turbine units)	9.7	7.1–8.7	3.7–7.5	2.5–6.3	7.1–8.7	4–7.5	3.3–6.3
CCGT	20.5	22.1–30.1	23.6–33.9	22.7–38.2	22.2–24.7	24–29.3	26.5–33.5
Gas turbine plants and diesel power plants	13.5	14.3–17.2	13.5–18.1	13.5–19.1	14.3–15.4	13.5–18.1	13.5–19.1
Nuclear power plants	9.9	9.3–10.8	8.5–12.5	2.1–18.3	9.3–10.5	8.5–10.9	2.1–10.7
Renewables (inclusive of hydroelectric power plants)	16.6	18.2–24.3	17.1–30	15.7–31.2	18.2–20.1	17.1–20.8	15.7–21.1
Distributed generation (gas)	0	0–0.3	0–0.7	0–1.4	0–0.3	0–0.7	0–1.4

Note The numbers correspond to the lower and upper bound values of the projection range respectively
Source Calculated by the author based on [22]

2040 and equal to 34–38 (cfr. 28% under the reference case being the most likely one).

The analysis of the projection range of the energy sector development, accompanied by the risk assessment of facilities that actually make up the range, is instrumental in making projections more defensible and facilitates the delineation of the window of opportunities and critical conditions. The very same aim is pursued by the development of the methods that are used for solving key problems that arise in the process of making projections as covered further in this book.

References

1. Gabor D, Colombo U (1978) Beyond the age of waste: a report to the Club of Rome. Pergamon Press, Oxford, p 264
2. Voropai NI (ed) (2010) Energy systems analysis: the Siberian Energy Institute/Energy Systems Institute schools of thought in hindsight. Nauka, Novosibirsk, p 686
3. Lagerev AV (2014) A dynamic territorial and production model for the development of scenarios of a mutually aligned development of Russia's energy industry as concerns its federal subjects. Izvestiia RAN. Energetika 4:26–32 (In Russian)
4. Kononov YD (2009) The MIDL macroeconomic model. Methods and models of projections of energy-economy interactions. Nauka, Novosibirsk, pp 143–146 (In Russian)
5. Kononov YD (1981) Energy and economy: the challenge to transitioning to new energy sources. Nauka, Moscow, p 190 (In Russian)
6. (1981) Energy in a finite world: a global systems analysis. Ballinger Publishing Company, Cambridge, MA, p 825
7. Makarov AA (2003) The system analysis of the prospective development of the energy industry. Izvestiia RAN. Energetika 1:42–50 (In Russian)
8. U.S. Energy Information Administration. The National Energy Modeling System: An overview [Electronic Publication]/U.S. Energy Information Administration. Retrieved from: http://www.eia.gov/oiaf/aeo/overview/
9. Voss A, Schlenzig C, Reuter A Mesap-III. A tool for energy planning and environmental management: History and new developments [Electronic Publication]. Retrieved from: http://elib.uni-stuttgart.de/opus/volltexte/2013/8529/pdf/vos280.pdf
10. Makarov AA (ed) (2011) SCANER. Model and information system. Energy Research Institute, Moscow, p 72 (In Russian)
11. Melentiev LA (1979) Energy systems analysis: sketch for a theory and directions for development. Nauka, Moscow, p 414 (In Russian)
12. Sessions R (1950) How a 'difficult' composer gets that way. New York Times, p 59
13. Moiseev NN (1984) The scientific prediction: illusions and realities. Znanie, Sila, No. 2 (In Russian)
14. Saenko VV, Kurichev NK (2013) Six steps of energy strategizing (the case of the Energy Strategy 2035/2050). Energeticheskaia politika 2:35–46 (In Russian)
15. Rubinshtein AL (1975) Methodological problems of economic projections of scientific and technological advances. In: Vilensky MA (ed) Economic aspects of technological forecasting. Ekonomika, Moscow, p 222 (In Russian)
16. Makarov AA, Makarova AS, Khorshev AA (2011) Prospects of the development of nuclear power plants to the mid-twenty-first century. Energy Research Institute, Moscow, p 210 (In Russian)
17. Ministry of Energy of the Russian Federation. The agency for forecasting electric power energy balances. Scenario assumptions for the development of the electric power industry to the year

2030 [Electronic Publication]. Retrieved from: http://ranipool.ru/images/data/gallery/1_8337_ _usloviya_elektroenergetiki_na_period_do_2030_goda.pdf (In Russian)

18. Ministry of Energy of the Russian Federation. Until the year 2020, with the projection time frame extended to the year 2030. [Electronic Publication]. Retrieved from: http://www. minenergo.gov.ru (In Russian)

19. Ministry of Energy of the Russian Federation. Energy strategy of Russian to the year 2035. Key takeaways [Electronic Publication]. Retrieved from: http://www.energystrategy.ru/projects/ docs/OP_ES-2035.doc (In Russian)

20. U.S. Energy Information Administration. International Energy Outlook 2013 [Electronic Publication]. Retrieved from: http://www.eia.gov/forecasts/ieo/

21. U.S. Energy Information Administration. Annual Energy Outlook 2013 [Electronic Publication]. Retrieved from: http://www.eia.gov/forecasts/aeo/

22. U.S. Energy Information Administration. Annual Energy Outlook 2014 [Electronic Publication]. Retrieved from: http://www.eia.gov/forecasts/aeo/

Chapter 3
Methods of Projecting Price and Demand for Energy Carriers

3.1 The Evolution Undergone by Methods of Energy Consumption Projections

There is a reasonably wide array of methods applied to the analysis and projection of demand for energy carriers. Among the methods that are most widely applied to making energy consumption projections that go up to 15 years into the future are the direct counting method and its variations. The essence of the method is to isolate a limited number of the most energy-intensive production activities (types of products and services), carry out an in-depth analysis of their energy efficiency, and estimate prospective energy demand based on projections of production volumes and energy intensity dynamics.

Merging the direct counting method with oftentimes quite sophisticated econometric models (that is, regression equations) saw its international heyday back in the 1970s and 1980s. A case in point is the MEDEE (Model for Long-term Energy Demand Evaluation) simulation model [1] that enables the evaluation of the impact of such factors as the make-up and growth rates of industrial production, the quality of life, the energy conservation policy for individual sectors and the like on energy consumption. The demand for energy is calculated for production and tertiary sectors of the economy.

The 1990s saw the arrival of systems of models for projections of energy consumption in the USA and Europe. PRIMES [2], developed to support the European Commission activities in impact assessments and analysis of policy options, is one such system of models. The system is used as a tool to analyze an energy policy as it interacts with energy technologies. The PRIMES models enable the calculation of energy balances (both yearly balances and long-term balances), demand for energy for specific energy carriers and economic sectors, energy carriers prices, CO_2 emissions, etc. Key variables of the model are balanced out as below: demand for energy carriers is a function of the price, production equals consumption, price is inversely related to production volumes of energy carriers. The energy consumption module

© Springer Nature Switzerland AG 2020
Y. D. Kononov, *Long-term Modeled Projections of the Energy Sector*,
Springer Geophysics, https://doi.org/10.1007/978-3-030-30533-8_3

calculates the demand for energy for residential and commercial sectors, agriculture, industrial production, transportation, as well as the energy sector industries. A more comprehensive overview of energy consumption models developed abroad is presented in [3].

In Russia, energy consumption projections have long been dominated by a number of the direct counting method variations. Here, much consideration was given to the assessment of a potential decrease in the energy intensity rates due to the introduction of new technologies and higher levels of energy conservation.

E.g., JSC Energosetproekt [4] developed a procedure for projecting the demand for electric energy with the break down by federal subjects of the Russian Federation and local subsidiary companies of the then OAO RAO UES (PJSC Unified Energy System of Russia) for the five-year period as based on the two methods: aggregated specific performance indicators (ASPI) and an econometric method.

The essence of the ASPI method is to calculate energy intensity values specific to economic sectors and industries (both in monetary terms and fixed prices) based on the published reported data. The above energy intensity values are then propagated all over the projection time frame which yields the so called basic projection, where the demand for electricity in each economic sector is assumed to be proportionately related to the growth rates of production and provision of services (in the case of the residential sector the demand is proportionate to the population). This is followed by modifying the initial projection in line with related inter- and intra-industry structural shifts and the potential energy conservation measures as well as the increased efficiency of electrification.

The Melentiev Energy Systems Institute of the Siberian Branch of the Russian Academy of Sciences introduced the use of input-output models to account for the impacts of structural changes in the industrial production sector on the fuel and energy demand of the national economy. Their undeniable advantage is the possibility to analyze and account for indirect energy links on a par with the direct ones. Together they manifest themselves as coefficients of total costs of individual energy carriers. They convey the information on the additional demand for energy across the whole network of production links should the final consumption of any product increase by one unit. Calculating and using these coefficients are at the core of the method of integrated energy consumption for components of final non-industrial consumption as developed by Medvedeva [5]. The method was further elaborated at the Melentiev Energy Systems Institute where they developed an approach based on non-monetary input-output modeling that enabled the calculation of the demand not only for energy carriers but also for energy-intensive materials and services.

The methods reviewed above lack sophistication with respect to accounting for complex and changing interactions between energy consumption volumes and the conditions and the development level of the economy and the energy sector. Incorporating individual energy consumption models in the overall system of the energy sector development facilitates overcoming such deficiencies.

At the Energy Research Institute of the Russian Academy of Sciences, this approach was implemented by means of simulation models [6]. Its workflow is outlined in Fig. 3.1. The EDFS (Energy Demand Forecasting System) model and information system was developed based on such models and supporting databases.

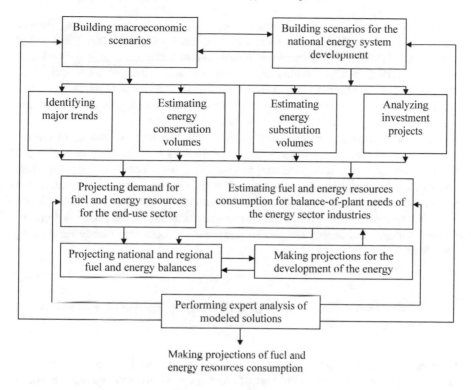

Fig. 3.1 The procedure developed at the Energy Research Institute of the Russian Academy of Sciences for projecting energy consumption at the national and regional levels. *Source* [6]

A similar approach is being developed at the Melentiev Energy Systems Institute. To this end, the emphasis is made on energy consumption in individual regions and federal districts of Russia [7].

The above models and model systems are used for projections with the time frame of up to 20 years. Longer-term projections are performed based on key indicators of the quality of life and specific consumption of energy required for each of the indicators to meet the corresponding projected value [8].

In the case of extreme long-term projections, it is permissible to extrapolate the most sustainable global trends data mined from statistical information alongside available medium-term projections. Such trends include the following ones: the decrease in cross-country variation in terms of the GDP energy intensity and per capita electricity consumption, as well as the elasticity of energy consumption with respect to the GDP.

Alignment of energy consumption projections with population and economic projections is at the core of the original approach applied to making projections of the world and Russian energy industry development out to 2040 [9]. To this end, first, for each of the 67 country groups as per the UN data on projections of the population growth, projections based on per capita energy consumption, the per capita GDP, and the GDP energy intensity are aligned with each other. Since either of the methods yields interval projections that set constraints on acceptable variation within the trends, an optimization problem is solved to arrive at projection estimates. The problem is essentially to identify two projection ranges for such trends that differ the least. Then, the total of country-specific projections is mutually adjusted in line with an independent world projection of energy consumption.

Further development of projection methods for long-term dynamics of energy consumption is accomplished by way of accounting for anticipated structural changes in the industrial sector, transportation, and the lifestyles adopted by the population. To this end, the model accounts for a potential increase in energy conservation on a par with the impact of the cost of energy carriers on its possible and economically feasible implementation. The calculation and applications of the price elasticity of demand for fuel and energy are gaining increasingly more importance for making projections

3.2 Long-Term Projections of a Probable Range of Fuel Prices

Long-term projections (those that go over 10–15 years into the future) of fuel and electricity prices are an essential part of all strategies and development programs of the energy sector as well as the national and regional economies. They provide insights on competitiveness of various energy companies, new fuel deposits, energy intensive product types, and serve as an important reference point when making investment decisions.

Development of exports and the market framework in the energy sector, reforming of the electric power sector and the gas industry all contribute to the growing complexity of energy-economy interactions, increase uncertainty of prospective cost estimates, and urge to continually improve the methods employed for making long-term projections. Projections of energy carriers prices per se also require regular updates and detailing.

A price projection should be linked to the energy sector and economic development scenarios. To this end, it is essential to account for the increased effect of the cost of energy carriers on economic growth rates and energy consumption. In theory, this implies solving an optimization problem of the mutually conditioned development of the national and regional economy and the energy industry, while accounting for the effects of price mechanisms. However, highly uncertain conditions, requirements and links, in our opinion, make it unfeasible to apply elaborate model systems with a single objective function to long-term projections of the price dynamics of energy carriers. A more realistic approach that matches the current level of expertise and available resources is an incremental approach that treats a price projection as a problem of its own and iteratively loops it together with the problems of projections of the economy, energy consumption, and the energy sector development.

We propose a procedure for calculating probable fuel and electric energy price dynamics as shown in Fig. 3.2. Its salient feature is its simulation of competition on energy markets and the treatment of price dynamics as a rightward-extending cone of their probabilistic estimates. To this end, as the upper boundary of fuel prices on Russian energy markets we use equilibrium prices that provide the return equal to that when using world prices in a given region (while accounting for actual possibilities of increasing exports). Such prices are equal to export prices net of transportation rates, transit fees for transportation across the territory of third-party countries, and customs charges. In some regions, the gas price competitive with the price of locally mined or imported coals can serve as a reference point for the upper bound of the gas price range.

Our analysis of global trends shows that oil and gas prices on world energy markets follow the prices of oil and petroleum products, lagging 6–10 months behind them. The cost of coal is usually 70–85% lower than the cost of oil of equivalent calorific value, while the cost of the pipeline gas is 25–35% lower than the cost of oil of equivalent calorific value. The cost of liquefied natural gas (LNG), the demand for which has enjoyed an increase by a factor of 20 over the last 25 years, is twice that

Fig. 3.2 The procedure for making projections of fuel and electricity price dynamics

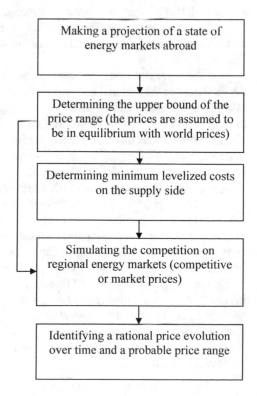

of the pipeline gas. Its price (inclusive of regasification costs) is only marginally different from that of oil, however, it is expected to reach the pipeline gas price.

The discrepancy between coal and gas prices on global energy markets has been increasing due to higher requirements for fuel quality and attempts at lowering the level of environmental pollution and emissions of carbon dioxide, and other so-called greenhouse gases. In the case of power plants, the development of high-performance combined cycle gas turbines also contributes to higher competitiveness of gas. According to the U.S. Department of Energy projections [10], the difference between gas and coal prices in the electric power industry is expected to increase to $US 68–75 per tce by 2020 (back in 1980, it amounted to $US 40 per tce), while the nowadays representative coal-to-gas price ratio of 1:1.8 for power plants will reach 1:(1.9–2.2) in 2020–2030.

In Russia, price ratios for various energy carriers are somewhat different: in 2005, the ratio of average cost of a ton of coal equivalent for coal, natural gas, and fuel oil for power plants was 1:0.9:4.3; in 2014 the same ratio was 1:1.4:3.7. The gas to coal price ratio is assumed to increase to 1.8 by 2020.

The lower bound of the price range is defined as minimum supply prices that enable the market entry by individual energy companies (levelized costs). For existing facilities, the above price value shall be sufficient to cover annual expenses, tax payments, and to ensure minimum profit required for normal operation. In the case of newly built facilities or a developing company, the levelized costs should also include the investment component. The latter shall ensure paying back interest-bearing debt and generating acceptable annual average return on investment during a given time frame.

It is the levelized costs that have to be used as an optimality criterion for the development of the national and regional energy sector before the actual projections take place, because their use ensures a well-grounded selection of likely competitors on regional coal, gas, petroleum product markets and cross-regional energy links. They are higher than discounted costs that are common in optimization models. The latter are a sum of the production costs and specific capital investments multiplied by the performance factor (rate of return).

A dedicated simulation model (INTAR) was developed at the Melentiev Energy Systems Institute of the Siberian Branch of the Russian Academy of Sciences for the calculation of levelized costs [11]. The model accounts for a diverse range of potential investment sources (depreciation, profits, loans, shares of stock), as well as the changing and asynchronous nature of the process of investing, paying back interest-bearing debt, annual expenses, and profits.

Comparing period averages of levelized costs helps to get an overview of competitiveness of various deposits and energy companies and enables making a selection of them to further detail the probable price dynamics and simulate a given energy market. The upper boundary of this dynamics range is defined by market prices. They serve as the basis for the valuation of investment projects (with price elasticity of demand factored in).

3.3 Methods and Results of Quantitative Assessment of the Price Elasticity of Demand for Energy Carriers

Established international practices provide evidence supportive of a significant dependency between the demand for energy carriers and the dynamics of energy intensity of specific industries and the economy as a whole, on the one side, and changes in fuel and energy prices on the other side. Their growth first triggers the substitution of one energy carrier for another and then leads to transition to energy conservation technologies (that is, one production factor is replaced by another: energy is substituted for capital investment and labor), and finally results in the replacement of energy-intensive products and services by less energy-intensive ones. As this takes place the response of the economy to a significant rise in price of energy carriers spans over long periods of time: straightforward switching between energy carriers takes up to 2–3 years, transition to new technologies requires up to 5–8 years, while structural changes in production and end-use consumption happen only in 8–10 years after the initial price change. This response can be captured as short-term and long-term elasticity, that is a percentage change in demand for a given energy carrier when its cost increases by 1%.

Elasticity values are usually determined by econometric models of the following type:

$$D_i = F(Y, P_i, P_j),$$

where F—generally, a logarithmic function; D_i—consumption of the ith energy carrier; Y—gross output or income; P_i—the price of energy carrier i; P_j—the price of a substitute energy carrier.

Price coefficients in these models built as based on published reports reflect the price elasticity of energy consumption.

It should be point out, that the type of the employed model and the set of its variables have a notable impact on the elasticity value. This statement is illustrated by Table 3.1 as well as by the findings obtained back in 1992 during the implementation of one of the projects hosted by Energy Modeling Forum, an organization that was responsible for the review of complex energy problems and was funded by the U.S. Department of Energy. Within the framework of that project an option of increasing the cost of all energy carriers by 25% was considered alongside the reference case of the U.S. energy systems development for years 1991–2010 (under the assumption of its 1% annual growth). The calculations performed by means of six independent models resulted in a notable discrepancy between estimates of the decrease in energy consumption by 2010 as compared to the reference case: 2–6% for the tertiary sector, 4–8% for the industrial sector, and 1–8% for transportation.

Table 3.2 also provides insights regarding the price elasticity of demand for the industrial and residential sectors (with the adaptation to the price growth factored in). The average values for coefficients of short- and long-term elasticity shown in

Table 3.1 The coefficient values of the price elasticity of demand for energy carriers arrived at by various econometric models for years 1970 to 1984

Country	The number of models	Elasticity	
		Short-term	Long-term
USA	2	−0.138 ÷ −0.141	−0.513 ÷ −0.517
Germany	5	−0.151 ÷ −0.485	−0.206 ÷ −1.170
Japan	5	−0.290 ÷ −0.551	−0.473 ÷ −1.080
France	2	−0.154 ÷ −0.448	−0.410 ÷ −0.448
UK	3	−0.01 ÷ −0.166	−0.1 ÷ −0.21
Canada	2	−0.39 ÷ −0.62	−1.07 ÷ −1.1

Source [12]

Table 3.2 The values of the price elasticity assumed for projections of energy markets development in Europe

Consumers	Gas	Petroleum products	Coal	Electricity
Short-term elasticity				
Residential sector	−0.22	−0.21	−0.19	−0.32
Industrial sector	−0.27	−0.20	−0.19	−0.20
Long-term elasticity				
Residential sector	−0.68	−0.89	−0.72	−0.64
Industrial sector	−1.12	−0.83	−0.86	−0.99

Source [13]

the table were applied to a study of potential consequences that radical deregulation of the European energy market might bring about.

The increase of the gas share in energy balances of many countries (including Russia) and a significant volatility of its cost attract keen interest to the estimates of elasticity of demand for it.

Our analysis of gas consumption in 30 countries from 1967 to 1998 reveals the tendency toward the non-linear growth of the price elasticity of demand for gas as the per capita GDP grows (Fig. 3.3). According to Tables 3.1 and 3.3, the short-term price elasticity of demand for gas in the USA and Europe does not exceed −0.30. However, its long-term value grows by 1.5–3 times in the tertiary sector and by 3–4 times in the industrial sector. World Bank averages (published in 2008) for the long-term price elasticity of the demand for gas: −1.35—for the industrial sector, and − 0.56—for the residential sector [14].

Studies of quantitative assessment of coefficients of the price elasticity of demand for energy carriers that were published internationally prove that their values drastically change over time and depend strongly on specific conditions and the local context of energy and economy development processes in various countries. That said, the studies enable drawing certain qualitative conclusions that are instrumental

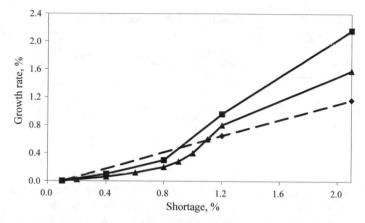

Fig. 3.3 The price elasticity of demand for gas as dependent on the per capita GDP. Legend: 1—Turkey, 2—Russia, 3—Romania, 4—Poland, 5—Greece, 6—Mexico, 7—Brazil, 8—Hungary, 9—Taiwan, 10—Australia, 11—Canada, 12—Germany, 13—UK, 14—Ireland, 15—The Netherlands, 16—Austria, 17—Belgium, 18—USA. *Source* Based on [15]

Table 3.3 The price elasticity of demand for gas in the USA (average values)

Consumers	Short-term	Long-term
Industrial sector	−0.24	−0.67
Population	0.24	−0.41
Commercial sector	−0.29	−0.40

Source [13–15]

in calculating the potential response to changes in the fuel and energy cost on the part of consumers:

- the price elasticity of demand increases as the economy becomes more market-driven and is growing alongside the growth of per capita GDP;
- the long-term elasticity (adaptation of the consumer over a longer time period to a fuel and energy price increase) is several times that of the short-term elasticity;
- coefficients of the price elasticity of demand for energy carriers in the tertiary sector are lower and less volatile than those in the industrial sector. The price elasticity of demand for fuel for power plants shows the most variation over time and across countries;
- when assessing a potential response to the quality of a given energy carrier, one should account for price changes for competing energy carriers (that is, for the cross-elasticity).

The above price elasticity coefficient estimates published abroad were arrived at by employing econometric models that require quite extended time series of reported data and reflect the past experience. It should be pointed out, that there is an emerging trend of research papers (see e.g. [16] and [17]) that advocate building econometric models and calculating the price elasticity of demand for energy carriers based on

Table 3.4 The price elasticity of demand for the tertiary sector in the USA

Sector	Energy carrier	Short-term			Long-term
		Year 1	Year 2	Year 3	Year 25
Residential	Electricity	−0.12	−0.21	−0.24	−0.40
	Gas	−0.08	−0.14	−0.17	−0.28
Commercial	Electricity	−0.12	−0.20	−0.25	−0.82
	Gas	−0.14	−0.24	−0.29	−0.45

Source [17]

processing the results of projections of the energy industry development under various price dynamics scenarios. This is a viable option given that projections are grounded in optimization models that juxtapose economic performance of the application of various energy carriers by various consumer groups.

The elasticity coefficient values shown in Table 3.4 were calculated by way of matching energy consumption under the reference case of the U.S. energy industry development against the scenarios that assume electricity and gas prices that are twice as high.

In Russia, as market mechanisms in the economy and energy industry mature, the response to changes in the cost of energy carriers on the part of various consumer groups will grow more and more tangible, while the numeric values of coefficients of the price elasticity of demand will approach those representative of the developed countries. Nowadays, available statistic data are insufficient for obtaining the values of coefficients of the price elasticity of demand for fuel and energy of any reasonable reliability to be suitable to be used for projections. Furthermore, somewhat idiosyncratic Russian realities lend themselves to the application of the elasticity estimates published abroad as rough estimates only. Therefore, the calculations of changes in demand for energy carriers should be based on direct comparison of economic performance of using various types of energy carriers by various consumer groups and account for social and environmental criteria and requirements.

3.4   Incremental Multi-stage Projections of Price and Demand on Regional Energy Markets[1]

A salient feature of our approach (see Fig. 3.4) is that it combines estimates of the price elasticity of demand for energy carriers with the optimization of the energy and of fuel supply of a region. To this end, the elasticity is calculated separately for each group of consumers of a given region (thermal power plants, boiler houses, industrial and domestic units, transportation), while accounting for the so called "consumer's

---

[1]The section is co-authored with Elena V. Galperova and Dmitry Yu. Kononov.

Fig. 3.4 Stages of
projections of the state (price
and demand) of regional
energy markets

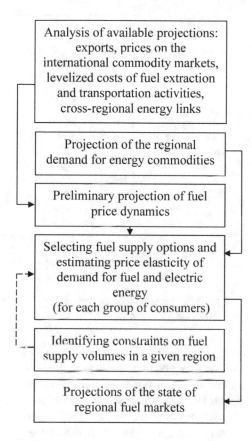

gain", that is the impact of a fuel type and its quality on technical and economic performance indicators of the consumer.

Hence the problem to be solved can be stated as follows: to find a solution that meets a given demand for energy carriers in the region under assumed conditions that is most efficient for a given group of consumers. In doing so, all key input data are given as intervals with various probability density functions governing their distribution.

To decrease the number of iterations and properly account for the ambiguity of the input data it is reasonable to provide a more in-depth treatment of those aggregated regions that feature the highest uncertainty of conditions of the electricity and fuel supply within a given time frame.

The choice of the fuel for projections of regional energy sectors is interrelated with the choice of the composition, location and technologies of new capacity additions. Therefore, the problem of estimating the demand for a given fuel by, for example, newly built power plants or boiler houses in a given region should be solved simultaneously with the optimization of the make-up of the generating capacity. Methods for solving such problems are well-established. Problems arise, however,

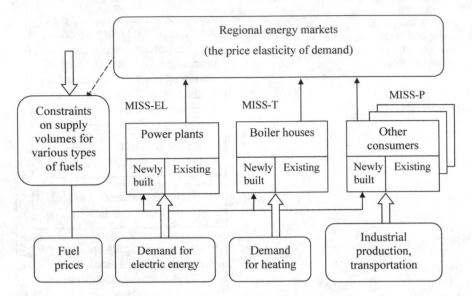

Fig. 3.5 Modeling and software system for projections of the price elasticity of demand for fuel on regional energy markets

when the amount of input data given as intervals is huge, while there is no unanimously accepted hypothesis as to the probability law that governs the distribution of values within such intervals. Furthermore, it is required to select the solution most robust to changes in assumed conditions (prices, demand, technologies, etc.) out of all solutions obtained by deterministic optimization models.

An approach to deal with the above difficulties when solving the problem of choosing options of fuel supply, reasonable and feasible new capacity additions in energy supply systems under interval uncertainty is implemented in the MISS (Monte Carlo Implementation of Static Stochastic Modeling) group of models that each target a specific group of consumers. They are integrated in a unified computing system (see Fig. 3.5).[2]

One of the features unique to this system of models is its application of optimization methods alongside statistical experiments (Monte Carlo methods). The former serves for choosing a rational mix of fuel supply of consumers, while the latter accounts for the uncertainty of future conditions. Another distinctive feature of the MISS models is that initial technical and economic performance data, prices, and other variables are all interval estimates.

When generating possible combinations of input data that reflect possible conditions of financing and operation of a given project, we rely on the Beta distribution probability density function:

$$F_x(a, b, \alpha, \beta) = (x - a)^{\alpha-1}(b - x)^{\beta-1}/B(a, b, \alpha, \beta),$$

---

[2]The software implementation of the MISS models was developed by Vladimir N. Tyrtyshny.

where $B(a, b, \alpha, \beta) = \int\limits_{a}^{b} B(a, b, \alpha, \beta) = (x-a)^{\alpha-1}(b-x)^{\beta-1}dx$; a, b—boundaries of the uncertainty range; α, $\beta > 0$—numeric parameters that define a type of the probability distribution within the uncertainty range. By varying parameters α and β it is possible to generate random variables of the most diverse types of probability distributions ranging from uniform to normal to lognormal to exponential, etc.

The simulation procedure is designed so that the random choice of values from within predefined probability distributions respects known or assumed correlations between variables.

In the case of calculating the demand for fuel by newly built power plants within a given time frame in a specific region or an energy system, key variables of interest in a simplified model (MISS-EL) are: capacity of newly built power plants, annual electric energy outputs, consumption of various fuel types. In doing so, the model accounts for constraints (as defined by the upper and lower bounds of interval estimates): on potential new capacity additions for each type of power plants, supplies of specific fuel types to a given region and their prices, electric energy exports and imports. The demand for electricity given as intervals is to be met in the most efficient way possible (e.g., based on the minimum electric energy cost criterion).

Results of simulations get aggregated, dependencies between variables of interest and changes in fuel supply conditions are identified, alongside the most efficient and robust solutions as well as probabilistic characteristics of key variables of interest. An important outcome of the investigation of various scenarios is arriving at probabilistic price elasticity of demand for the fuel of this group of consumers.

By generalizing calculation results performed by various groups of consumers (with mandatory individual treatment of newly built power plants and major boiler houses), one can determine the total demand for individual types of fuels in this region as a function of their prices.

Modeled calculations proved a significant impact that the kind of input data uncertainty exerts on this dependency. This is, in particular, demonstrated by the results obtained under assumed conditions of the fuel supply for newly built boiler houses in various regions of the country (see Table 3.5) and newly built power plants (see Table 3.6).

When making a projection of the state of regional energy markets, maximum available volumes of supply of individual energy carriers to this region within a

Table 3.5 Coefficients of price elasticity of demand for gas for newly built boiler houses as a function of a region and the kind of input data uncertainty

Region	Interval uncertainty	Normal distribution	Deterministic solution
European Russia	−0.67	−0.61	−0.13
Eastern Siberia	−0.50	−0.49	−0.30
Far East (South)	−0.58	−0.41	−0.20

Note Results of calculations under the conditions assumed for 2020
Source [18]

Table 3.6 The effect of the interval width and assumed probability distribution of input data and fuel costs on the fuel demand by power plants

Performance indicators	Units of measurement	Interval uncertainty (uncertainty proper)		Normal distribution of data	
		Lower price fuel	Higher price fuel	Lower price fuel	Higher price fuel
Fuel prices	$US per tce				
Gas	-"-	120–135	158–175	132	165
Coal	-"-	73–79	81–91	74	87
Demand for fuel	Thousands tce				
Gas	-"-	1850–2060	1420–1640	2030	1540
Coal	-"-	2750–2860	3140–3250	2760	3210
The share of new capacity additions	%				
Gas-fired power plants	-"-	44–47	34–38	47	36
Coal-fired power plants	-"-	53–55	61–62	53	62

Note Results of modeled calculations based on the MISS-EL model for assumed conditions of annual average capacity additions of power plants in European Russia in 2020–2025

given time frame are a crucial bit of input information. First and above all, this is relevant to available gas volumes that are transported to the region via main gas pipelines.

An isolated treatment of a geographic area does not allow to properly account for the constraints on the supply of gas or other types of fuel to this region from the deposits of the national importance or those that are export-oriented. These constraints are defined by the overall fuel balance, a potential to increase fuel extraction and transportation, economic performance and investment risks of various options as perceived by suppliers (energy companies).

The problem of estimating possible and efficient cross-regional energy links and supplies of fuel and energy to regions is solved, first and foremost, at the level of the national energy sector. Given the inevitable ambiguity of solutions, in order to provide a more detailed treatment of constraints on available resources of fuels from specific deposits in a given region in some cases it is reasonable to apply specialized stochastic optimization models of suppliers. In such models a rational strategy of supply is defined given the projection range of potential demand identified at the previous step and prices in various regions. The problem is solved so as to achieve the maximum profit for the supplier company given interval estimates of production volumes, production and transportation costs, and the price elasticity of demand in regions.

After adjusting constraints on supply volumes of a specific fuel to a given region so that they fit more detailed data as the latter become available, stochastic modeled calculations for consumers are reiterated.

The above approach to incremental multi-stage projections of the state of regional energy markets awaits further development. Of top priority is finding a rational way of input data evaluation and properly accounting for price and demand projections at various hierarchical levels. It seems reasonable to incorporate the "Regional energy markets" module into the overall procedure of making projections for the energy sector (see Fig. 2.4). To this end, the output information of this model generated during iterative calculations and fed into models of various levels has to contain estimates of the price elasticity of regional demand for fuel and energy.

References

1. Chateau B, Lapillonne B (1982) The MEDEE approach: analysis and long-term forecasting of final energy demand of country. Energy modelling studies and conservation: Proceedings of a seminar of the United Nations Economics Commission for Europe, Washington D.C., 24–28 March 1980. Elsevier, pp 57–67
2. Mantzos L, Capros P The PRIMES. Version 2. Energy System Model: Design and features [Electronic Publication]. Retrieved from: http://www.e3mlab.ntua.gr/manuals/PRIMESld.pdf
3. Kononov YD, Galperova EV, Kononov DY (2009) Methods and models for projections of energy-economy interactions. Nauka, Novosibirsk, p 178 (In Russian)
4. Kogan YM (2006) Current problems of forecasting the demand for electricity: the report delivered at the « Economic problems of the energy sectors » open workshop, the 59th meeting held on March 29, 2005. Izdatelstvo INP, Moscow p 34 (In Russian)
5. Medvedeva EA (1994) Technological paradigms and energy consumption. Siberian Energy Institute of the Siberian Branch of the Russian Academy of Sciences, Irkutsk, p 250 (In Russian)
6. Filippov SP (2010) Forecasting energy consumption based on a system of adaptive simulation models. Izvestiia RAN. Energetika 4:41–55 (In Russian)
7. Galperova EV, Kononov YD, Mazurova OV (2008) Price-driven projections of the regional demand for energy carriers. Region 3:207–219 (In Russian)
8. Makarov AA, Makarova AS, Khorshev AA (2011) Prospects of the development of nuclear power plants to the mid-twenty-first century. Energy Research Institute, Moscow, p 210 (In Russian)
9. Makarov AA (ed) (2013) The projection of the energy industry development in Russia and the world to the year 2040. Energy Research Institute of the Russian Academy of Sciences—Analytical Center for the Government of the Russian Federation, Moscow, p 110 (In Russian)
10. U.S. Energy Information Administration. U.S. Annual Energy Outlook 2013 [Electronic Publication]. Retrieved from: http://www.eia.gov/forecasts/aeo/
11. Kononov YD, Kononov DY (2006) New requirements and approaches to long-term projections of electricity prices. Izvestiia RAN. Energetika 3:3–9 (In Russian)
12. Komigama R (2010) Projecting long-term natural gas demand as a function of price and income elasticities. Research Report. IIASA, Laxenburg, p 10
13. Aune FR, Brekke KA, Golombek R et al (2008) LIBEMOD 2000—LIBeralisation MODel for European energy markets: a technical description. Working paper 1/2008. Ragnar Frisch Centre for Economic Research, Oslo
14. Huntington HG (2007) Industrial natural gas consumption in the United States: an empirical model for evaluating future trends, Energy Econ 29(4):743–759
15. Lady GM (2010) Evaluating long term forecasts, Energy Econ 32(2):450–457

16. Wade S. Price responsiveness in the AEO2003 NEMS residential and commercial buildings sector models [Electronic Publication]. Retrieved from: http://www.eia.gov/oiaf/analysispaper/elasticity/
17. Ellis J The effects of fossil fuel subsidy reforms: a review of modeling and empirical studies [Electronic Publication]. Retrieved from: http://www.iisd.org/gsi/sites/default/files/effects_ffs.pdf
18. Galperova EVA (2013) Theoretically-grounded approach to estimating energy consumption under uncertainty. Region 3:212–218 (In Russian)

Chapter 4
Assessing and Factoring in Conditions and Barriers that Limit the Projection Range of Prospective Development of the Energy Sector

When developing projections for the energy sector one has to consider multiple cases of its development as harmonized with the national economic growth scenarios, volumes and the composition of exports and imports, the expected change in pricing and tax policy, and the governance framework. Obviously, the growth rates of the energy sector industries as well as the speed of their structural changes that can be accomplished with each time period are limited, so the arbitrary high demand for energy carriers cannot be met.

High capital intensity and the inertia inherent in the energy sector industries urge us to pay due attention to their capacity to ensure the accelerated development of the Russian economy, the growth of energy consumption, and a significant increase in exports of energy resources. A possible bottleneck can manifest itself as a lack of time or that of materials, funding, and labor necessary for new capacity additions not in the energy sector itself but in its supporting industries. Timely availability of these resources for the development of the energy industry and related industries is conditioned by the comprehensive effect of multiple factors inclusive of energy carriers pricing, a state of the energy markets, government policies, the availability and efficiency of imports of products and technologies.

We have to admit that to date there are no acceptable and universally recognized approaches to deal with deciding on whether a given option of long-term development of the energy sector can or cannot be adopted. The same applies to the comprehensive quantitative assessment of the barriers that put constraints under certain conditions on growth rates and structural changes of energy systems.

Below, we make an attempt at developing a provisional concept of barriers in the energy sector, highlighting major problems and challenges that are part of the research, and providing tentative solutions.

© Springer Nature Switzerland AG 2020
Y. D. Kononov, *Long-term Modeled Projections of the Energy Sector*,
Springer Geophysics, https://doi.org/10.1007/978-3-030-30533-8_4

4.1   Taxonomy of Constraints and Barriers of the Energy Sector Development

Barriers are defined as existing bottlenecks that can potentially hinder the development of energy systems under the conditions anticipated for a given time period. They are identified by way of juxtaposing the energy sector development requirements and the capacity to meet them. Quantitative estimates of barriers can serve as constraints in economic and mathematical models that are employed for making projections.

The higher the hierarchical level the more complex the systems representative thereof grow and the more significant their inherent inertia is. Furthermore, the more challenging it becomes to overcome the barriers that appear when accelerating the development rates and when the structure of such systems has to undergo changes.

All barriers can be subdivided into the following major groups based on their origin and the way they manifest themselves when it comes to making projections:

- Temporal barriers: constraints on the timing of adding new capacity in the energy sector that are caused by a lack of time for their engineering, construction, and development.
- Financial (investment) barriers: a shortage of funding required for capital expenditures.
- Resource barriers: a shortage of required materials, labor, and natural resources that are available for development.
- Technological barriers: inability to provide new technologies and equipment within a given time period to ensure the required development.
- Price barriers: prices that are either unacceptably low for individual producers or unacceptably high for consumers hence hindering their economic well-being.
- Short-term sales (marketing) barriers: constraints on the demand for specific energy resources on domestic and international markets.
- Environmental barriers: rigid requirements for environmental protection.
- Political and organizational barriers: constraints put on the energy sector development by requirements for energy and national security and government policies.
- Barriers that make correct estimates and efficient strategic decisions difficult because of the uncertainty of future conditions and related risks.

Barriers also might include the state of such system-defining qualities as flexibility and reliability that fail to comply with the requirements.

Barriers can also be grouped based on other criteria: importance (major, minor); rigidity (unsurmountable, surmountable under certain conditions); timing of possible occurrence and how long effect endures (short-term, medium-term, long-term), structure (immediate, indirect) and complexity (simple, compound).

Compound (complex) barriers include those determined by the properties of a given system. First and above all this applies to the properties of inertia and adaptivity.

In case when judged by these properties a system does not meet new requirements, this can hinder proper development of the system.

The list of barriers can be extended and modified depending on a problem and a given hierarchical level. In doing so, it makes sense to distinguish between constraints and barriers that are exogenous and endogenous to a given hierarchical level (system) (see Table 4.1).

Of all endogenous constraints one of the most significant ones are *temporal barriers* conditioned by the development inertia of energy systems.

High capital intensity of the fuel industry and electric power industry, their direct production links with the machine industry, the iron and steel industry, and other production industries, with the transportation industry and the construction sector, as well as a significant amount of time required for the construction of energy facilities, the development of the infrastructure and look-ahead development of linked production facilities (see Fig. 4.1) all contribute to high inertia levels of the energy sector.

The latter manifests itself, in particular, as an inability to sharply increase production volumes within a short time period, to change the composition of facilities in individual industries of the energy sector as well as the make-up of the national energy balance.

The inertia property and methods of its assessment and treatment in projections are covered in the dedicated section below.

Apart from temporal barriers, major obstacles to accelerating the development, modernization of production facilities, and new capacity additions include *investment and resource* barriers.

The assessment of required capital expenditures as well as specialized equipment and materials, and human resources of required qualification levels is necessary for building development scenarios of individual companies and entire industries alike. Such assessment presents no particular difficulties given the availability of design documentation, specifications and guidelines, the use of past experience, and similar projects. It is way more complex to assess the possibility of providing required financial and material resources, especially in the case of projections for a remote time horizon and the subject of projections being not the development of a single company but of the whole industry. Under such conditions, even approximate quantitative assessment of investment and resource barriers gains a particular significance.

Investment barriers are related to *price barriers* as financial resources required for making investments to a significant degree are a function of profits.

On energy markets, prices are shaped under the impact of competition between suppliers, hence balancing off demand and supply. In the case of individual energy companies and supplying companies the market price can serve as a barrier if it proves lower than levelized costs (that is minimum acceptable supply prices that mark the threshold value below which the production and delivery of fuel and energy are deemed sub-economic).

The price barrier for a given energy carrier for the consumer arises in the case when its expected price proves unacceptable based on economic or other reasons and there is an alternative solution available.

Table 4.1 Constraints on the energy sector development at various hierarchical levels

Hierarchical level	Internal constraints	External constraints
1. Companies, enterprises	Available production capacity (assets, technologies, labor, reserves). Financial resources. Performance of projects and their investment risks. Time required for construction and modernization.	Demand for company's products, market prices, export and import opportunities, competition, infrastructural constraints, directives.
2. Systems of individual industries	The scale and the required time for potential development of resource fields and new capacity additions. Available capital investments. Availability and throughput capacity of major transportation links. Constraints on the development of individual companies (as applied to new capacity additions by regions).	Volumes and patterns of demand for products of a given industry, potential for its exports, market prices, directives, assignments, and regulations.
3. Regional energy sectors	Proven resources of fuel and energy resources, required time and amount of new capacity additions in the electric power industry and the fuel industry within the region.	Demand for fuel and energy. prices. Cross-regional energy links. Environmental and social requirements
4. The national energy sector	Production volumes and development times of major centers of fuel production, potential for new capacity additions in the electric power industry and the fuel industry.	Demand for energy carriers, boundaries on potential exports and imports of the energy sector products, prices on international and domestic energy markets, indicators of national security and national security, limits on CO_2 emissions.

Note Key constraints factored in the development, the assessment, and the choice of long-term development options

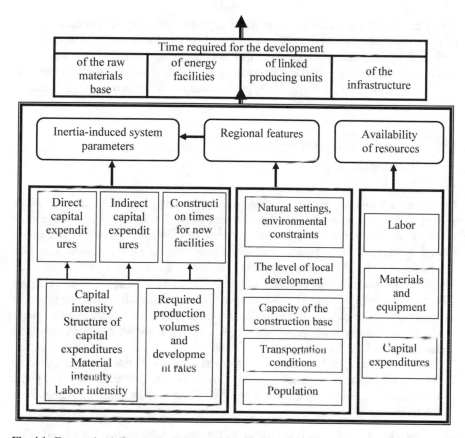

Fig. 4.1 Factors that influence the duration of the development of new centers of fuel production

Insufficient demand on domestic and international markets within a given time frame can become a major obstacle (*a barrier conditioned by demand*) to the development of new centers of fuel production or the construction of new power plants.

The assessment of minimum demand for fuel and energy required for normal operation of energy facilities that are either existing or being built or being planned is a problem of much import in the study of potential barriers in the energy sector. Of even more importance and way more complexity is the problem of estimating the upper bound of the acceptable growth of demand. Exceeding this value may result in either a shortage of energy carriers within a given time period or an excessive increase in fuel and energy prices.

Higher uncertainty of future demand and prices for energy carriers as well as the values of other input parameters result in challenges to be faced when estimating the return of planned projects, especially in the cases of the valuation of development options for the systems of individual industries and regions. The larger the uncertainty is the higher investment risks are and the lower the probability of providing required funding and other resources for planned new capacity additions are. The combination of factors that are unfavorable for investors such as uncertainty, risks, insufficiently high performance, and time constraints may become a major *barrier* for *new capacity additions* in the energy sector industries. A valid quantitative assessment of this compound barrier is one of top priority problems to tackle when making projections.

4.2 Inertia-Induced Barriers to the Energy Sector Development

Inertia is an inherent property of large developing systems. To be deemed well-grounded no projection can do away with accounting for this property. The importance of investigations into potential quantitative manifestations of inertia under changing conditions of the energy industry development is reflected in the works by academician Lev A. Melentiev. He defined inertia as an ability of systems to resist development understood as external and internal stimuli that target to change the previously projected trajectory [1], and he treated inertia as bundled with the property of flexibility as constraining the latter. By flexibility, he meant the ability of a system to change its strategy at a required rate to ensure normal development and operation under potential disturbances.

Economic inertia may arguably be captured in terms of the efforts required to change a development trajectory (growth rates, structure, composition) of a given economic system. When applied to energy systems such efforts manifest themselves as the following indicators:

(1) total (direct and indirect) capital expenditures or total costs of labor and other resources in the national economy spent on production and consumption of an additional unit (in mln. tce) of a given energy resource with factoring in of the costs of development of linked industries, as well as the production and social infrastructure;
(2) time required to implement all such capital expenditures and all supporting measures (inclusive of design and survey work, provision of necessary facilities on site, etc.);
(3) maximum incremental increase in the production of a given energy resource that can be achieved in n years per each billion rubles of additional capital expenditures.

The latter indicator is compound in nature and is derived from the former two ones. Quantitative assessment of all of the above indicators and their presentation in the form of functional dependencies enable us to compare and rank various centers of fuel production and alternative energy sources based on their inertia levels.

To capture relative inertia of development of entire industries one can apply two indicators that complement each other: (a) minimum time required to increase incremental growth rates of the products produced by the industry by 1 percentage point or its share in the gross output of the industry by 1%; (b) additional expenditures on the part of the resources of the national economy that are required to achieve this.

If one is to treat capital expenditures as a key factor that leads to changes in the structure and development rates, then total capital intensity can be understood by analogy with physics as the body weight, which is telling of the inertia of the system. The higher the value of this indicator is, the higher the inertia is.

The intrinsic inertia of the energy sector and its subsystems is made up of the inertia of new centers of fuel and energy production and newly built energy facilities and technologies the development of which defines it prospective make-up.

Obviously, inertia indicators for the development of new centers of fuel production, in addition to the features specific to a given energy production, is also influenced by region-specific features such as natural and climatic conditions, the level of development of the area, its distance from potential suppliers and consumers, capacity of the construction base, labor balance of the district, etc. The less favorable regional conditions are, the higher required direct and coupled capital expenditures for the implementation of a program are.

Of all external conditions that influence the inertia level of the energy sector, a key role is played by the level of development of linked industries and the time required to produce equipment and materials to boost production, conversion, and transportation of energy resources. The implementation of large-scale projects in the electric power industry and the development of new centers of fuel production may require beforehand new capacity additions in energy-related machine industry, the iron and steel industry, the construction industry, and other industries and production units that support the energy sector. The higher the growth rate of the energy sector industries is, the higher, in general, the number of production facilities involved in backing it and the more important the role of remote adjacent links are. Our calculations prove that additional (indirect) capital expenditures in such production facilities and their lead time grow in a non-linear way as the energy sector development accelerates (see Fig. 4.2).

Constraints on feasible timings of new capacity additions in the energy sector are conditioned first and above all by the time required for engineering and design work and construction of key facilities (temporal barriers of level I).

Their assessment is facilitated by the availability of investment projects developed by energy companies. In the case of their absence one has to rely on expert judgement alongside available data on similar projects and normative time limits for construction. Empirically-derived dependencies that link time limits for construction (t) to the amount of required capital expenditures (I) may also prove useful:

$$t = \alpha I^{\beta}.$$

One such dependency was obtained by Roald M. Merkin [2] by processing a large amount of data on major American enterprises with the following values assigned to empirically defined ratios: $\alpha = 2, 4$, $\beta = 0, 2$ (if t is measured in months, while I is in dollars in thousands).

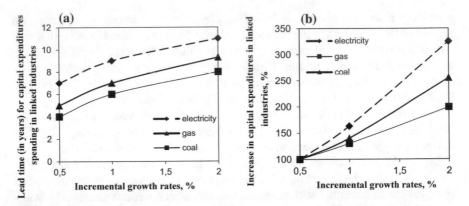

Fig. 4.2 a–b The general pattern of the effect an increase in incremental growth rates for the production of energy carriers has on an increase in time and capital expenditures required for the adequate development of linked industries. *Note* Not inclusive of additional imports of equipment and materials

The non-linear growth of the time required for construction as directly related to the increase in required capital expenditures is furthermore proved by Fig. 4.3 plotted based on Russian and American data on capital expenditures (in 2009 prices) and time limits for construction of power plants of various types.

The energy sector development inertia is related to the inertia of the entire economy. The higher its ability to make maneuvers with financial and labor resources, to change the structure of its industries and promptly respond to changing situation and state of international markets is, the easier it is to ensure required changes within the energy sector and its linked industries alike. On the other hand, the energy industry contributes to the flexibility of the economy by lowering its inertia.

It follows from the above that a valid treatment of the inertia inherent in the energy sector development is to be performed as looped together with the entire economy. To

Fig. 4.3 The effect that power plants construction costs have on the time it takes to put them into operation. Legend: ▲—in the USA, ♦—in Russia. *Sources* Calculated by the author based on [3] and [4]

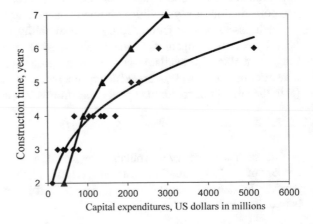

Fig. 4.4 The information flow diagram of the IMPAKT-2 model

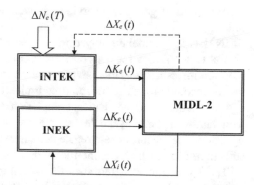

this end, an analysis of the inertia inherent in the development of individual centers of fuel and energy production cannot do away with due consideration of its role in the national energy sector.

Published research [5–12] on inertia and flexibility of energy systems saw its heyday in the USSR back in the 1980s. A similar line of research was conducted at the International Institute of Applied Systems Analysis (IIASA), Austria [13, 14]. In recent years there has been a strongly felt need to deepen and reconsider the notion of the inertia of energy systems and the importance of temporal barriers under new conditions of the development of the national energy sector and economy.

4.3 Methods and Results of the Study of Temporal Barriers and Inertia of Energy Systems[1]

The input-output IMPAKT model [10] was developed as early as back in 1975 to enable the study of the inertia property. The model formalized the process of retrograde unfolding (i.e. from the future back to the present) of links of various levels from a given energy facility to production facilities that ensure its development. A shortcoming of this model was that it failed to account for expected economic conditions. To a certain extent, this issue is addressed by the IMPAKT-2 system of models developed at the Melentiev Energy Systems Institute of the Siberian Branch of the Russian Academy of Sciences (see Fig. 4.4). Its software implementation is made up of three modules: INTEK, INEK, and MIDL-2

MIDL-2 is an updated version of MIDL, a well-established dynamic optimization macroeconomic model [15] of the input-output type. The new version of this model accounts for the annual dynamics of economic development unlike the earlier version that was limited to the temporal resolution of five-year periods within the 30-year long time frame. The model depicts mutual production links between 29 branches of the national economy and between production and non-production industries via the

---

[1]The section is co-authored by Dmitry Yu. Kononov.

consumption of goods and services, as well as investment links and export-import links.

INTEK is a model employed for the assessment of required dynamics of direct capital expenditures for the construction of new energy facilities (inclusive of the required infrastructure) that are part of a given development option. To this end, the model accounts for standardized time limits for construction and the distribution of equipment costs and mechanical completion costs over the span of all years starting from land-use planning and development to commissioning and start-up of the facility. The model enables to aggregate certain variables and presents them in a way that suits their use in a macroeconomic model.

The INEK model determines the dynamics of capital expenditures for linked industries and production facilities. Requirements for their development rates are determined upon obtaining a corresponding solution from the MIDL-2 model.

IMPAKT-2 modeled calculations are performed as per the following iterative procedure under the assumption that the reference case of the development of the economy and the energy sector is pre-defined:

1. MIDL-2 is calibrated so as to match the reference case.
2. The INTEK model then calculates the dynamics of new capital expenditures based on a pre-defined case of new capacity additions in the energy sector or one of its subsystems. The respective additional demand for equipment and construction and installation work are then fed into MIDL-2 together with the data on the required increase in the energy sector production volumes in year T.
3. The solution provided by MIDL-2 and its comparative analysis against the reference case enables to determine the rate of required additional development of linked industries (first-level linking), as well as the corresponding increase in demand for energy carriers.
4. These data ($\Delta X_i(t)$) are fed into the INEK model that calculates additional capital expenditures required in linked industries.
5. In the case of a significant increase in the fuel and energy demand in linked industries, INEK is used to detail the capital expenditures required by the energy sector.
6. The results of calculations of steps 4 and 5 are then transferred to MIDL-2 in order to detail the dynamics of the additional development required of the economy relative to the reference case and to identify more remote linking levels.

By varying the product imports of various industries in MIDL-2, one can assess their impact on the inertia level, that is to say, the required look-ahead development of linked industries and its scale.

The following IMPAKT-2 model calculation results for a simplified case study provide an overview of potential impacts on the scale and time limits of required additional development of various branches of the economy, on the increase in the electric energy production and equipment imports. The scenario of the development of Russian economy and energy industry that assumes reaching the 1.5 trillion kWh level of electric energy production by 2030 was used as the reference case. As its alternatives we also studied the growth of the electric energy production by additional 5% and 15% in 2026 to 2030.

Our calculations suggest that total capital expenditures of linked industries given no equipment and materials imports can exceed direct investments in the electric power industry by 1.2 times under additional growth rates of the electric energy production by further 5%, and by 1.35 times under growth rates of 15%. The share of the gas industry and other fuel industries in the overall composition of additional capital expenditures for linked industries, the transportation industry and the telecommunications is 15–22%, the construction industry accounts for 12–19%, the machine industry is 2–6%, the iron and steel industry is 2–3%, while the rest of the industries make up 11–12% altogether.

It is essential that a significant share of capital expenditures, the required increase in the demand for industrial products and services take place in the years prior to the surge in the electricity production. Based on our calculations, the increase in the demand for industrial products, construction and installation work, the transportation and tertiary services manifests itself 5–10 years prior the required 5% increase in the electricity production and by 10–15 years and more in the case when such incremental growth increases up to 15%.

Imports of required equipment contribute to a notable decrease in the level of inertia by eliminating remote linking levels. This is illustrated by the data in Table 4.2 that reflects the results of calculations for the case of an incremental increase in the electricity production by 15% subject to the constraint that makes 60% of additional

Table 4.2 The effect of 60% imports of equipment on lowering the demand for products of linked industries and bundled capital expenditures

Industries	Decrease, %	
	Gross output	Capital expenditures
Machine industry	79	75
Construction industry	7	3
Oil and refinery industry	21	15
Gas industry	9	10
Coal industry	8	1
Iron and steel industry	54	50
Chemical industry	49	43
Construction materials industry	11	3
Transportation industry	29	20
Other industries	25	20
Total	34	17

Note Relative to the hypothetical case of the maximum development of the electric power industry with no imports of energy equipment
Source Calculated by the author for the case of the incremental 15% growth of the electricity production volumes relative to the reference case

demand for the machine industry products (relative to the reference case) satisfied by imports in the time period from 2020 to 2030. The calculations suggest that the additional demand for the products of all linked industries decreases by 34%. Here, most drastic (by 50% and more) is the decrease in the required production of ferrous and non-ferrous metals and chemical products. Freight turnover and demand for petroleum products fall by 21–29% (not inclusive of gas pipelines). Required production output by the machine industry decreases over the entire period more than the volumes of additional imports of energy equipment due to the decrease in the demand for linked capital expenditures. In this case the look-ahead development at the beginning of the investment process decreases as follows: by 8–12 years in the machine industry, by 5–10 years in the iron and steel industry, up to 5 years in other linked industries.

4.4 Approaches to the Assessment of Investment Barriers in Projections of the Energy Sector Development

When developing programs for individual industries alongside alternative options of the energy sector development, it is crucial to properly estimate and account for potential investment barriers that set constraints on new capacity additions and pose a threat of a shortage of energy resources within a given time frame.

In optimization models that are employed to identify possible and feasible options of the energy sector development such barriers can be captured as constraints put on the amount of capital expenditures. Obviously, it is impossible to arrive at the values that are of any consequence doing away with the scenarios of social and economic development published by the government as they elucidate the investment policy adhered to at the national level. In practical work, when making assumptions regarding the energy sector development and scenarios of economic development, one has to rely on the trends that manifest themselves as changes in the share of the energy sector in total capital expenditures incurred by the production sector while factoring in the assumed GDP and gross product growth rates as well as anticipated changes in the share of the energy industry in the total output.

The highly hypothetical nature of direct estimates of the constraints on the amount of capital expenditures incurred in the energy sector urges to invent indirect methods of estimating investment barriers. One of these indirect methods is to simulate possible behavior patterns of prospective investors. The latter depends on the anticipated performance and the level of risk of projects and programs that make up any scenario of the energy sector and its industry development.

Table 4.3 Risk coefficient values for various categories of oil reserves

Reserves category	Risk coefficient
Proven reserves	
Proven developed reserves	
Operating lifetime > 5 years	0.7
Operating lifetime < 5 years	0.65
Proven undeveloped	0.5
Probable	0.25
Possible	0.10

Note Estimates of the net present values (calculated at the 10% discount rates not adjusted for inflation and increases in oil and gas prices) are adjusted as per the above coefficient values
Source [19]

4.4.1 Methods of Assessment of Investment Project Risks[2]

Factoring in uncertainty and risks is an indispensable part of the valuation of investment projects (that is to say, their commercial feasibility). A project is deemed robust if under all scenarios it proves profitable and commercially feasible while possible unfavorable consequences can be mitigated by the measures provided by the project's organizational and economic structure itself [16].

There are several methods to estimate the robustness of a project under uncertainty [17, 18]. They include contingency calculations of the project value by varying the values of technical and economic performance indicators and anticipated operating conditions of the project. To this end, one applies statistic methods to process calculation results and then arrives at probabilistic risk estimates based on the values of dispersion and coefficients of variation for variables of interest.

The methods of adjusting the discount rate to accommodate risks gained widespread acceptance. For example, the U.S. Energy Information Administration assumes a discount rate of 10% with 7% of this value accounting for the risk premium. This discount rate is applied to valuation of investment projects undertaken by electric utilities. Back in the 1970s and 1980s, the risk premium was smaller and amounted to 4%.

Calculating the net present value (*NPV*) at the risk-free discount rate and then multiplying the resulting NPV by a pre-defined risk coefficient can be considered a variation on the risk-adjusted discount rate approach (see Table 4.3).

In [20], the authors propose to account for the effects of risk on the project returns by adjusting the risk-free values of the internal rate of return (*IRR*) as follows:

$$NPV' = NPV - \beta \sum_{i=0}^{I} X_i,$$

---

[2]The subsection is co-authored by Vadim I. Loktionov.

where X—the cardinal value of losses caused by the risk eventuating at time i, i—the share of non-mitigated losses.

Methods of calculating the points that are critical for a project's success, including the break-even point analysis, have been adopted worldwide. The above method, while being the most straightforward of the approaches to project valuation, provides financial input information on a par with other methods that inform the valuation of investment projects. To this end, one calculates the minimum allowable (that is, critical) sales (or production) level at which the project still breaks even. The lower this level is, the more probable it is that the project will survive under shrinking target markets, and hence the lower the risk for the investor is.

It should be pointed out, that the availability of universally recognized metrics of investment project valuation notwithstanding, there are no "rigorous" methods of quantitative risk assessment. In most cases, risks are assessed based on expert judgements.

A case in point is that during the preliminary work on The General Scheme (Master Plan) of the electric power industry development under the projected price dynamics assumption for years 2009–2020, power generation companies raised the issue of risk assessment for the investment projects they proposed. Each type of risk was assigned a certain weight, which was followed by estimates of individual and integral risk measures based on expert judgements. Out of the six considered types of risk the highest weight was assigned to the uncertainty of financing sources (21%) and the shortage of fuel supply (19%). Projects that scored over 66% were regarded as very risky. They amounted to 15% and 20% of all projects presented by energy companies for the entire time frame and the 2010–2015 period respectively.

It is evident that the choice of a risk assessment method for a certain project depends on its scale and importance as well as on the nature (quality) of input data.

Large-scale investment projects are known to have extended duration of the construction project and the lifetime and are able to influence the input data assumed for the project valuation via feedback links. Oftentimes, these variables can be factored in only as interval estimates without further details on the probability law that governs the distribution of values within the intervals. Of utmost uncertainty are usually the input data (prices, demand, capital intensity) for large-scale projects related to exports of energy sources, imports of equipment, and those dependent on the volatile state of international commodities markets. In such case, it may by reasonable to apply the multi-stage risk assessment procedure presented in [21].

As the first step of the procedure, one decides on the risk premium (Δr_i) for each risk-contributing factor separately. To this end, one shall first calculate the net present value (NPV) when the variable of interest takes its minimum and maximum values at the same discount rate. Then, when calculating the NPV under the most favorable scenario, one is to calculate the discount rate that would made the NPV equal to the NPV under the worst-case scenario. The difference between the above two discount rates is telling of the risk that corresponds to a given risk-contributing factor (Δr_i).

When calculating the Δr_i value, other variables of interest that influence the NPV, are assigned their expected values (in the case when their probability density functions are known) or are calculated based on the well-known Hurwicz's decision rule (for interval variables):

$$f = \lambda f_{max} + (1-\lambda) f_{min},$$

where f_{max}, f_{min}—are the maximum and the minimum values respectively as assumed by the variable; λ—the coefficient of pessimism/optimism, which generally takes the slightly pessimistic value of 0.3 when one lacks the relevant probabilistic data.

As the second step, one calculates the integral risk performance of the project (R) that is determined by all variables. It is essentially a weighted-average sum of all Δr_i:

$$R = \sum \gamma_i \Delta r_i,$$

where γ_i is the contribution of the ith factor to the NPV.

As the third step of the calculation procedure, the value of risk measure R is compared against the project performance measured, for example, as its return on investment, that is the ratio of the NPV to the amount of capital expenditures (PI). For each competing project one calculates the PR (profit—risk) ratio that is indicative of the return per each unit of risk for a given project:

$$PR = \frac{PI}{R}.$$

This ratio can serve as an important indicator for ranking projects based on their economic performance under interval input data uncertainty. Its shortcoming, however, is that it does not capture various degrees of risk aversion that a potential investor may entertain. Generally, when dealing with large expected profits, the risk aversion decreases. Accordingly, the maximum acceptable value of PR grows incommensurably large. If there is information available on risk aversion of a potential investor, then at the final stage of calculations it is reasonable to plot the curve showing the maximum allowable risk values as a function of expected returns on investment. Coordinates of the point telling of risk and return values for each of the options under study, relative to this curve, inform on their competitiveness while accounting for investor's risk aversion. Options that have such point as more remote from the risk aversion curve given the same values of PI shall be deemed more preferable. This corresponds to the maximum value of the variable of interest that can be defined as the subjective risk measure:

$$CR = \frac{R'(PI) - R(PI)}{PI},$$

where $R(PI)$ is the value R takes for a given project that yields the return rate PI; $R'(PI)$ is the indifference curve value of R at the return rate PI.

The $R'(PI)$ value is indicative of the maximum allowable risk value for the investor at a given return rate PI.

The above approach was applied to the problem of ranking alternative development options for the much promising Kovykta gas condensate field, for exports and other options of making us of the gas [22].

4.4.2 Approaches to Risk and Economic Performance Valuation of Options of Prospective Development of Energy Companies and the Energy Sector Industries

Major energy companies and the energy sector industries alike cannot limit themselves to individual projects when performing risk assessment. The entire "life cycle" curve of an industry (that is, the total of dynamics of all variables that are telling of its current financial standing) has to be subject to risk analysis over a long time span of prospective development that exceeds the typical lifetime of an individual investment project. When managing the development of a company, the risk assessment of a decrease in its market value, ratings, and business failure proves more important than the assessment of profitability of individual projects.

An attempt at quantitative assessment and ranking of individual investment projects alongside options of the entire energy sector industries development is presented in Table 4.4. The risk map (based on a 10-point scale) presented there was composed under various levels of demand on domestic and global gas markets as part of the preparation of the General Scheme (Master Plan) of the gas industry development through 2030 [23]. It should be pointed out that within this work, as is the case with other similar publications issued by the majority of planning and design institutions, investment and other risks are subject to expert judgements.

Theoretical developments in the field of investment risk assessment and decision making in the energy sector under uncertainty [24, 25], as elaborated as they were back in the USSR era, never saw their proper application in the planned economy of that time. Likewise, it is only in the last 15–20 years that the risk assessment of the energy industry development received due attention, which contributed to the growing sophistication of models and the strategy of their application to projections.

International practices of electric utilities include the use of mean-variance analysis, a method based on the modern portfolio theory. This is a method for the assessment of expected return and risk of an array of construction projects of various facilities (that is, several power plants) under anticipated changes in demand, production volumes, and prices on a given market.

In [26], the authors present a two-stage approach to the rational portfolio (structure) selection for generating capacity to meet predefined prospective electricity demand (with the price elasticity of demand factored in) and the load curve under uncertainty.

Table 4.4 Risk map for the gas industry development

Risks	Projected demand		
	domestic market - high, global market - moderate	domestic market - low, global market - high	domestic market - moderate, global market - low
A significant excess of price growth rates for basic materials and resources in the gas industry over gas price growth rates	4.8	6.2	3.7
Low efficiency of geological prospecting in newly developed regions	0.3	1.5	0.2
Failure to provide new technologies and material and technical resources in a timely manner	6.5	7.0	2.0
Sunk investments in the development of production facilities in the case of an unplanned decrease in gas production volumes	1.5	1.2	0.9
Domestic gas demand in Russia lagging behind the planned growth rates	9.8	2.0	6.0
Decrease in the growth rates of demand for the Russian gas abroad	5.7	8.1	4.8
Unfavorable state of export markets	4.2	7.3	3.7
Slowing down of the process of deregulation of the domestic gas market, sustaining the policy of artificial lowering of gas prices	6.9	1.9	3.1
Risk levels (based on the 10-point scale)			
high	*moderate*	*low*	

Source Based on [23]

The purpose of the first stage is to identify the change in demand given the change of electricity generation costs under a given elasticity coefficient. A change in electricity generation costs is defined as the difference between its value under the reference case and the values it takes as a result of Monte Carlo simulations for each portfolio of capacities under possible changes in controlled variables. At the second stage, which serves as the key stage, generation costs are calculated for each combination of production capacities alongside the probability density function parameters of such combination, which is telling of investment risks. The calculation is also a Monte Carlo simulation. During the second stage, the effect of elasticity of demand on the load profile is factored in.

In Russia, as is also the case abroad, risk assessment methods that are tailored to the programs of long-term development of energy companies and energy industries have been gaining increasingly more attention. Recently, we have seen the emergence of indigenous developments in Russia as well. One of them is an approach to risk analysis advocated by the Melentiev Energy Research Institute of the Siberian

Branch of the Russian Academy of Sciences, as applied to well-defined production and investment programs of the development of individual industries [27]. The system of simulations is governed by a Monte Carlo procedure. The optimal solutions obtained for each random combination of risk factors with relevant data on the performance of the industry development program are processed statistically: intervals of possible values and estimates of mean values, etc. are calculated for each performance indicator. Then, the most preferable strategies are chosen based on a quantitative analysis of indicators of the industry performance within a given time frame, so as to assess the risk inherent in them. The approach outlined above was employed at the Energy Research Institute of the Russian Academy of Sciences to perform a risk analysis of the Gazprom program of expanding the Unified Gas System eastward.

References

1. Melentiev LA (1979) Energy systems analysis: sketch for a theory and directions for development. Nauka, Moscow, p 414 (In Russian)
2. Merkin RM (1978) Economic problems of cutting down the construction time. Ekonomika, Moscow, p 175 (In Russian)
3. Howells M et al (2010) Incorporating macroeconomic feedback into an energy systems model using an IO approach: evaluating the rebound effect in the Korean electricity system. Energy Policy 38, Incorporating macroeconomic feedback into an energy systems model using an IO approach 6:2700–2728
4. Makarov AA, Makarova AS, Khorshev AA (2011) Prospects of the development of nuclear power plants to the mid-twenty-first century. Energy Research Institute, Moscow, pp 210 (In Russian)
5. Kononov YD (1981) Energy and economy: the challenge to transitioning to new energy sources. Nauka, Moscow, p 190 (In Russian)
6. Smirnov VA (1983) Adaptation processes in the energy industry development: theoretical issues and analytical procedures. Adaptation process in the energy industry development. Nauka, Moscow, p 196 (In Russian)
7. Smirnov VA () Problems of enhancing the flexibility in the energy industry. Nauka, Moscow, p 191 (In Russian)
8. Deich IG (1985) Energy-economic trends in the development of production facilities. Nauka, Moscow, p 176 (In Russian)
9. Kononov YD (1981) External production links and inertia of the energy sector. Izvestiia SO AN SSSR. Seriia obshchstvennykh nauk 2:12–18 (In Russian)
10. Kononov YD, Korneev AG, Tkachenko VZ (1979) Modeling external production links of systems of individual industries. Ekonomika i matematicheskiye metody XV(5):969–977 (In Russian)
11. Kononov YD, Korneev AG, Tkachenko VZ (1988) Timelines of major energy programs. Siberian Energy Institute of the Siberian Branch of the Academy of Sciences of the Soviet Union, Irkutsk, p 75 (In Russian)
12. Karkhov AN (1990) Economic dynamics and the energy industry projections. Institute of Atomic Energy—5050/3, Moscow, p 39 (In Russian)
13. Energy in a finite world: a global systems analysis. Ballinger Publishing Company, Cambridge, MA, 1981, p 825
14. Kononov YD, Por A (1979) The economic impact model. RR-79-8. IIASA, Laxenburg, p 72

15. Kononov YD, Lyubimova EB, Tyrtyshny VN (1983) Issues of assessing macroeconomic consequences of long-term strategies of the energy industry development. Ekonomika i matematicheskiye metody 5:912–916 (In Russian)
16. Kossov VV, Livshits VN, Shakhnazarov AG (eds) (2000) Guidelines on the methods of the valuation of investment projects. Ekonomika, Moscow, p 421 (In Russian)
17. Livshits VN, Vilensky PL (2014) On common misunderstandings in the valuation of real investment projects. Ekonomika i matematicheskiye metody 1:3–23 (In Russian)
18. Tsvirkun AD (2008) The business plan. Investment analysis. Methods and software tools. Os'-89, Moscow, p 319 (In Russian)
19. Belkina EY, Dunaev VF (2012) A method of expert valuation of project risks as applied to research-and-development activities of an oil-and-gas company. Neft', gaz i biznes 1–2:12–16. (In Russian)
20. Tarasenko SR (2009) Projecting the value of an investment project with the variables capable of mitigating negative externalities factored in. Problemy prognozirovaniia 3:161–166 (In Russian)
21. Kononov YD, Loktionov VI (2010) Accounting for investment risks when ranking large-scale projects based on their economic value. Upravlenie riskami 1:48–51 (In Russian)
22. Loktionov VI (2010) Risk-informed ranking of alternative options of Kovykta gas development. Izvestiia Irkutskoy gosudarstvennoy ekonomicheskoi akademii (BGUEiP) 5:184–188 (In Russian)
23. The Ministry of Industry and Energy of the Russian Federation (2008) The general scheme (master plan) for the development of the gas industry until the year 2030 (draft). Moscow, p 145 (In Russian)
24. Lukyanov AS, Eskin VI, Shevchuk LM (1995) Quantitative risk assessment when choosing investment strategies in energy systems. Izvestiia RAN. Energetika 6:57–62 (In Russian)
25. Shevchuk LM, Lukianov AS, Kudryavtsev AA (2000) Risk analysis in the strategic planning problems for major energy companies. Izvestiia RAN. Energetika 2:52–64 (In Russian)
26. Vithayasrichareon P, MacGill IF (2012) A Monte Carlo based decision-support tool for assessing generation portfolios in future carbon constrained electricity industries. Energy Policy 41:374–392
27. Tarasov AE (2010) Approaches to risk mitigation during the eastward extension of the Unified Gas System of the Russian Federation. The energy industry of Russia in the 21st century: development strategy: facing eastward. Proceedings of the joint symposium. The Melentiev Energy Systems Institute of the Siberian Branch of the Russian Academy of Sciences, Irkutsk, p 699, pp 647–654 (In Russian)

Chapter 5
Quantitative Assessment of Strategic Threats and Energy Security Indicators

5.1 Overview of Energy Security Threats and Indicators

The amount of published research on energy security issues and attempts to measure it have been looming large in a wide range of countries. Integral to this process is the enormous diversity of opinions held by those who contribute to the energy security discourse. Their analysis (a survey of over 83 definitions) as presented in [1] proves that availability of energy resources is indispensable for the 99% of the energy security definitions. The great bulk of them also cover infrastructure development (72%) and energy prices (71%). In recent years, measuring of the following energy security facets have been taking an increasingly prominent role: environment (34%), societal effects (37%), governance, and energy efficiency (25%).

A conceptual-level definition of energy security has been suggested by the International Energy Agency (IEA, [2]), which reads as follows: "the uninterrupted availability of energy sources at an affordable price". To this end, they distinguish between the short-term energy security aspect and the long-term one. The former focuses on the ability of energy systems to properly respond to sudden changes in the energy supply-demand balance, while the long-term energy security mainly deals with timely investments in the energy sector to satisfy the future needs of the economy and meet the requirement for the environmental sustainability [2].

The choice of a definition of energy security proves to have a decisive impact on the tools applied to its measurement and monitoring. That said, virtually all of them are based on the application of indicators and indices. The former serve to quantify individual facets of energy security, while the latter are multi-faceted metrics, one of the major accomplishments of the indicator analysis technique.

In [3], the authors list as much as 320 simple indicators and 52 complex indicators employed by policymakers and academics in various countries to track national performance on energy security.

In Russia, the most comprehensive body of research on the indicator analysis technique and other energy security issues has to this day been published by the researchers at the Melentiev Energy Systems Institute of the Siberian Branch of the

© Springer Nature Switzerland AG 2020
Y. D. Kononov, *Long-term Modeled Projections of the Energy Sector*,
Springer Geophysics, https://doi.org/10.1007/978-3-030-30533-8_5

Russian Academy of Sciences (including three book-length presentations of this research field to their credit [4–6]). The Institute also served as a major contributor to the development of the Energy Security Doctrine of the Russian Federation [7]. The Doctrine defines energy security as the state of being protected against threats of shortages in satisfying the needs of citizens, the society, state institutions, and the economy with available energy resources of acceptable quality. The above definition is telling of the unique conditions that Russia's energy and economy have to operate and develop under. Unlike most other countries that 'outsource' their fuel supply, Russia is a net energy exporter. This well-established fact notwithstanding, Russia's vast territory knows no shortage of regions that lack energy self-sufficiency. That is why one has to give utmost consideration to the energy supply security not only at the national level but also at the level of individual regions. All of the above features cannot but have an effect on the composition and the importance of those indicators that are employed for the comprehensive assessment of the national energy security.

The analytic and forecasting framework proposed in [8] to deal with energy security performance blends the indicator analysis with modeling techniques. To this end, the set of all indicators is subdivided into three interrelated groups: (1) availability of production capacity and resources, (2) energy supply reliability, and (3) performance of fixed capital assets. The calculation of the integral energy security performance indicator based on the indicator analysis technique presupposes the following steps:

1. Selection of indicators, setting their thresholds, and normalization, that is qualitative performance scoring that makes the whole gamut of values boil down to acceptable, pre-critical, and critical states.
2. The assignment of importance weights to each indicator by means of interpolation and averaging out the opinions of third-party experts.
3. The aggregation of indicators that result in arriving at qualitative integral scoring of the overall national or regional energy security performance.

This performance is judged to be critically poor if the share of indicators ended up scored this way, relative to the total number of indicators, exceeds a predefined threshold value (for example, 40%).

Performance scoring of the national energy security is further elaborated by means of a dedicated energy sector optimization model and simulation models of energy sector industries [9]. This way, one explores an energy industry development scenario under possible shortages of energy resources on the consumer side, and then calculates extra costs to prevent it. Such costs serve as a basis for scoring energy security performance as acceptable, pre-critical, or critical.

Leveraging the analysis with a system of mathematical models to track energy security performance and assess the impact that ongoing energy and economic developments may have on the latter is a key defining feature of the above approach. In Russia, scholars and policy makers alike pay special attention to the analysis of performance and the monitoring of changes in the regional energy security [10]. The main takeaway from the indicator analysis is the qualitative assessment of the severity of threats of energy and fuel supply security that regions and the country as a whole will face. The energy security performance can be deemed critical or close to critical if numeric values of key indicators or aggregates thereof markedly differ from their predefined threshold values [11].

It is important to point out that energy security threats per se are defined, in general, based on expert judgements. The lack of satisfactory methods for quantitative assessment of strategic threats affects the performance of long-term projections.

Strategic threats are comprehensive in their nature and fraught with long-term and large-scale inhibition of the national economy growth rates [5]. Major strategic threats as identified in [5, 12] are listed in Fig. 5.1. Obviously, the list is non-exhaustive. It is possible and necessary to detail and amend it, given the changing external and internal developments in the energy industry both at the national and regional levels. To this end, it is essential to differentiate between permanent and temporary threats, primary and secondary threats, threats to the normal operation (sustainability) and development of the system.

One of key strategic threats is that of a capacity shortage: possible lagging behind of fuel production centers, electric power industry, and transportation infrastructure relative to the growing demand for fuel and energy. The shortage may result from the energy sector inertia, that is a lack of time required for the construction of energy facilities and for the development of infrastructure and linked industries and the

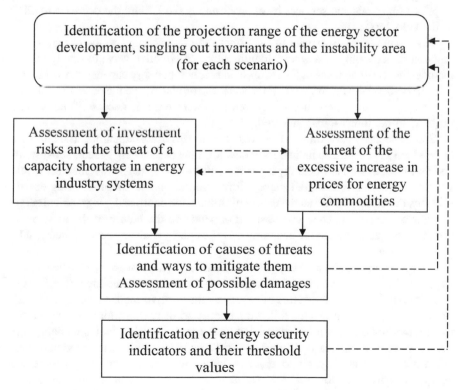

Fig. 5.1 Strategic threats to Russia's energy security and interactions between them. *Source* [12]

development of new territories. Another obstacle for the timely addition of required capacity may be resource constraints (financial, material, and labor ones).

The price dynamics for energy carriers that is unfavorable for the development of production and social sectors may become a major strategic threat that affects the national security. The reasons for the price growth may include the following: growing costs of commissioning and operating of newly developed centers of fuel and energy production, the state of international energy markets favorable for exports, erroneous pricing and tax policies, and other factors. The increase in the price of energy carriers encourages energy saving but this may prove to be insufficient to suppress inflation, to lower the competitiveness and the rate of return of certain energy intensive production facilities, and to sustain a desirable level of the quality of life of the population.

It should be pointed out, that considerable attention given to the assessment of the effect of the price factor on the energy strategy has its origins in the sharp increase in international oil prices in 1979–1980. Back in 1986, several universities and governmental agencies on the executive order by president Reagan performed an analysis of energy and national security that focused on economic losses to the oil shock. The major findings of this research program included, among others, that of the growth of international oil prices by 1% as capable of triggering the growth of prices for goods and services by approximately 0.69% and the decrease in GDP of the USA by 0.21%.

The quantitative assessment of the effect the price changes in the energy sector have on the economy as a whole, identification of effective ways to timely absorb the negative consequences of the increase in prices for energy carriers, the search for price trade-offs between interests of the state, the producer, and the consumers of fuel and energy are an important yet unsolved (to the extent it is necessary) theoretical problem of the forecast research methodology.

Energy security performance is assessed by means of its indicators. The latter are understood as generalized indicators of development of the energy sector subsystems and facilities, as well as of energy consumers that are selected so that their set would provide a reasonably comprehensive information on the status of ensuring energy security, the composition and the nature of threats, the depth and geographical scope of their coverage [5]. Their composition depends on the nature of the anticipated threats, features unique to the economy and the energy industry development, and a given time frame.

Energy security indicators enjoy much attention in the research published across the globe by major international agencies who focus on constructing and assessing quantitatively complex indicators of energy security performance [see e.g., 13–22]. For net importer countries, the following energy security indicators are deemed the most important ones: import dependence on suppler countries, fuel and energy mix diversity with respect to fuel types and geographical sources, international markets prices (most notably, oil prices), and depletion of resources. Many aggregated energy security indicators that are used internationally are constructed based on a modification of the well-known Shannon species diversity index. The International

Energy Agency utilizes the following aggregated energy security indicator to measure the risk of excessive dependence of a country on various fuel markets [13]:

$$ESI_{price} = \sum_f \left(\sum r_i S_{if} \right) C_f / TRES,$$

where S_{if}—the share of supplier i on market f; r_i—country-based political risk rating i; $C_f/TRES$—the share of fuel f in the overall energy supply.

Among other energy security indices used abroad, of most interest is the IMP_{tr} indicator that indicates the percentage share of the GDP the country is ready to spend to mitigate the risk of energy supply interruption:

$$IMP_{t,r} = Ai_{t,r}^{\alpha} C_{t,r}^{\beta} E_{t,r}^{\gamma}$$

where i—the share of the fuel in total imports; C—the share of the fuel in total energy consumption; E—energy intensity; A—a region-specific constant; α, β, γ—coefficients (equal to 1.1, 1.2, and 1.3 respectively) that reflect the non-linear risk growth.

An extensive list of energy security indicators suitable for the conditions specific to Russia is presented in [5]. They include the following: the ratio of the increment of known reserves of hydrocarbons to their production; the share of equipment that nears the end of its service life; the ratio of fixed capital assets subject to upgrading; the diversification ratio of the input side of the energy balance, the proportion of fuel exports relative to fuel production.

Just like energy security threats, the indicators lend themselves to being divided into those descriptive of the current state of energy security and those that are suitable for forward-looking estimates of strategic threats. The gravity of such threats is defined by judging the numeric values of energy security indicators against their threshold values. To this end, the following three performance scores of energy security are introduced (i.e. the three values assumed by indicators): acceptable, pre-critical, and critical [12].

Energy security serves as the basis of the national economic security and the key factor that defines its operation and development.

In 1998, The Ministry of Economy of the Russian Federation proposed 154 economic performance indicators. The Institute of Economics, Russian Academy of Sciences, advocated for the acceptance of 45 economic security indicators [23] with threshold values assigned to them, and they grouped them as follows:

- indicators that are telling of the overall economic potential and the ability of the economic system to undergo sustainable development (GDP volumes , gross industrial production volumes, gross yield of grain, the GDP share of expenditures on defense and research, etc.);

- indicators of the robustness of the financial system that warn about the likelihood of occurrence of critical situations in the public sector and the monetary sector (federal budget deficit, the share of the currency component as percentage of ruble-based money supply, public internal and external debt, etc.);
- indicators of the life quality of the population that are meant to define the limits that are, when violated, become threats to social and political stability (the unemployment rate, the share of individuals with revenues less than a living wage, the decile ratio, etc.);
- indicators that describe imports-exports security (the share of goods supplied as imports in the total volume of goods, etc.).

A number of methods have been applied to this day to assess and measure economic and energy security: scenario analysis and processing, optimization, expert judgements, game theory, fuzzy systems, multi-component statistical analysis, etc. That said, all the diversity of the above approaches notwithstanding, as of time of writing there are no universally recognized techniques of economic security performance assessment that would cover a comprehensive range of application domains.

Major methodological shortcomings of existing economic security indicator systems are as follows [24]: first, this is an empirical approach with its focus on statistical correlations at the cost of downplaying cause and effect links that exist in the economy; second, it is arbitrary in its selection of indicators and hence indicators get scattered as jigsaw puzzle pieces that provide no overall picture; third, there is a static treatment of economic phenomena irrespective of their dynamics, which is reflected in the proposed indicators and variables; fourth, the lack of any clues as to how to assess the damages (losses) that are inevitably done to the economic system under every deviation from the maximum permissible (threshold) indicator values.

These shortcomings hold true for energy security indicators as well. The importance of their quantitative assessment was well-recognized and considered in the source materials that formed the basis of the Energy Strategy of Russia (Table 5.1).

The indicator values listed in Table 5.1 are most likely to be treated not as threshold values but as target values based on expert judgements. There are no in the very least satisfying methods of quantitative assessment of threshold (critical) values of indicators for various time frames and various scenarios of economic and energy systems development.

An attempt at enhancing the validity of the approaches to quantitative assessment of strategic threats and energy security performance indicators for long-term projections of the energy sector development is presented in the following sections of this chapter.

Table 5.1 Energy security indicators

Indicator	Values specific to stages of the strategy implementation, %		
	Stage 1	Stage 2	Stage 3
Dynamics of domestic fuel and electricity prices for the industrial sector	less than or equal to the inflation rate		
Decrease in the average depreciation of fixed capital assets (as percentage of the 2010 values)	by 10%	by 10%	by 5%
The share of fuel and energy resources sector in the total volume Russia's exports	less than or equal to 60%	less than or equal to 55%	less than or equal to 45%
The share allocated to the APR countries in the total fuel and energy resource exports	not less than 17%	not less than 25%	not less than 28%
GDP energy intensity (as percentage of the 2010 values)	less than or equal to 76%	less than or equal to 65%	less than or equal to 50%
GDP power intensity (as percentage of the 2010 values)	less than or equal to 81%	less than or equal to 73%	less than or equal to 61%
Decrease of specific fuel consumption for electric energy generation (as percentage of the 2010 values)	not less than 10%	not less than 15%	not less than 18%
Ratio of the annual increment of primary fuel and energy resources reserves of various kinds to production volumes	not less than 1%	not less than 1%	not less than 1%

Source Energy Strategy of the Russian Federation to 2035 [25]

5.2   Quantitative Methods to Assess the Threat Posed by a Capacity Shortage

Quantitative assessment of the capacity shortage and other strategic-level threats presupposes the identification of the time when the threat can realize as well as the severity of the harm it can cause. The latter requirement implies the calculation of the following: the magnitude and the probability of a capacity shortage and potential harm due to it, as well as the costs to thwart the threat from realizing.

The energy sector optimization models that are usually employed to back up projections and balance out the demand and supply of energy carriers assume no capacity shortage. Contingency calculations based on such models are capable of providing at best an approximate estimate of changes in the development and the costs in a given system under the changes of constraints on new capacity additions.

The dedicated model developed at the Melentiev Energy Systems Institute of the Siberian Branch of the Russian Academy of Sciences to assess the performance of energy systems and the energy sector in the case of contingent events under various development scenarios better suits the purpose of the assessment of capacity shortages [26]. The model solves the problem of optimization of fuel and energy balances of regions of Russia (with the federal subjects of Russia treated on the one-by-one basis) under potential disturbances, which represents a classical linear programming problem in terms of its mathematical apparatus. Content-wise (that is, with respect to its energy-economic implications), the model is based on a well-established territorial-production model of the energy sector with individual modules for the electric power industry, the heat, gas, and coal supply, as well as the oil refinery and the fuel oil supply.

The objective function of the model is as follows:

$$(C, X) + \sum_{t=1}^{T} \left(r^t, g^t\right) \to \min$$

The first component of the above objective function accounts for costs related to the operation of the energy sector industries, the energy systems and subsystems that are part of it, as well as the capital expenditures spent on their development. Here, C is a specific cost vector for individual processes in the operation of energy facilities that are either operating or being reconstructed or being upgraded or being newly built. The second component is the losses due to a shortage of each fuel and energy type for each of the selected category t of the consumer. The magnitude (g^t) of the shortage of energy resources for the consumer of a given category is equal to the difference between the pre-defined maximum value and the desired value of consumption volumes for individual types of fuel and energy sources. The r^t vector is made up of the components that are somewhat tentatively to referred to as "specific losses". The difficulty of cost assessment of the actual magnitude of the damage due to a shortage can be overcome (if tentatively) by way of introducing a priority scale with respect to meeting the demand for individual energy carriers.

The problem of the assessment of possible contingent events in the energy sector and investigating the options of its development with respect to energy security performance calls for quite a detailed description of systems of individual industries with detailing of unique regional features (the model accounts for 90 geographic areas). In addition, the model solves the problem of the assessment of possible options of oil and gas supply to satisfy the demand of the consumer under contingent events (to this end, the model employs specialized flow models).

High dimensionality and the features unique to the above problem setting contributed to the choice of the steady-state rather than dynamic modeling that provides analytic treatment of appropriate energy security issues in a given year and under given conditions. Here, it is assumed that fuel and energy production capacity (its upper limit) is known alongside the maximum throughput capacity of major gas and oil pipelines and cross-regional power supply lines that are already in place.

The analysis of the threat of a possible shortage of regional energy supply without due consideration of its dynamics limits the choice of available means to mitigate it to those based on changes in reserves, transportation flows of fuel and energy, on a par with other measures that do not require considerable capital expenditures and do not have to be implemented well in advance. The above shortcoming of the above steady-state model proves more critical as the projection time frame extends further into the future. It can be mitigated by using the model as looped together with a dynamic yet more aggregated energy sector optimization model.

Additional gains in terms of identifying the significance of the capacity shortage threat can be expected if the workflow of the incremental analysis of such threat is extended to cover the investment risk assessment of key large-scale projects that are to be implemented as part of the systems of individual industries of the energy sector. Ruling out projects with unacceptably high risk for a potential investor from further consideration, introducing appropriate adjustments into the input data of the dynamic energy sector model, and performing new batch runs of calculations are 'the usual suspects' when it comes to assessing the potential mitigation of the shortage threat by means of timely structural changes in the energy sector under given conditions.

The probability and the severity of the capacity shortage threat should be assessed at the regional level as well. To this end, the problem can be treated as an analysis of possible risks of energy and fuel supply of a given territory under uncertainty. Obviously, both the input data that are required to solve the problems of this kind and the resulting risk estimates of energy supply options for individual regions have to be linked (harmonized) with overall projections of the national energy sector.

Given the above considerations, we propose the following two-stage approach to quantitative assessment of the strategic-level threat of a capacity shortage in the energy sector that accounts for the effect of the assumed economic development scenario and a certain hypothesis underpinning the energy carriers pricing policy to the extent they can influence these threats.

At the national level, most attention is given to the threats of insufficient development of the facilities of the national importance (large scale projects of hydrocarbon field development, export and cross-regional gas and oil pipelines, cross-system electric power transmission lines, etc.).

To this end, optimization models of the national energy sector serve as the key instrument for contingency calculations (under various assumptions of prices and future conditions). The composition of employed models, their properties (steady-state vs. dynamic), and the level of detailing (aggregation) depend on the projection time frame.

The capacity shortage risk is a function of the position a given facility takes within the projection range: the more different it is from the invariant solution and the less frequently the facility is part of optimal solutions, other things equal, the higher its associated investment risks are. At the regional level, we solve the problem of the assessment of strategic-level threats to their reliable energy supply.

When applied to the risk assessment of the regional electric energy supply in a given year (time period), the calculation may be carried out as per the following procedure:

1. The MISS-EL stochastic optimization model is set up with the objective to minimize the electric energy cost in a certain region under given demand for it. Key variables of interest are capacity of power plants to be added and the electricity price, while key constraints and conditions are as follows: demand for electric energy for all consumers in the region, its production by existing plants, possible exports and imports of electric energy, fuel prices, technical and economic performance indicators of power plants. All input data are given as intervals of their possible values with their respective probability density functions specified.

2. The simulation process for model calculations (hundreds of Monte Carlo tests) is performed so that a random sampling of a combination of input data does not violate known or assumed relations (correlations) between variables.

3. Test results are subject to the statistical analysis to estimate the probability of the inclusion in optimal solutions of each of the power plants (of a given capacity). Based on the above results, the risks for potential investors are identified with respect to the implementation of construction projects of individual plants. In general, the risk is equal to the ratio of the number of cases that have a given facility (construction project) included in an optimal solution to the total number of solutions (tests).

4. The most acceptable solution is identified for both new capacity additions (treated as an average of all tests or based on one of well-established criteria of decision-making under uncertainty—think Hurwicz's criterion and the like) and the corresponding average and market price of the electric energy.

5. The risks of the above solution (option) are assessed based on the investment risk of the plant that closes up the regional energy balance for capacity, and the average of the risk values for all the plants that are to be added.

These calculations enable finding the relative efficiency and risks of new capacity addition options in the region on a par with the demand of power plants for various fuel types under uncertain conditions. Furthermore, the model is solved for corresponding electric energy prices that are then used for making a rational choice of energy carriers for the industrial sector, the non-industrial sector, and the transportation sector.

Calculations of the comparative efficiency and risks of energy and fuel supply options for individual groups of consumers are performed by means of the modified stochastic MISS models detailed in Chap. 3.

Given the information obtained at the regional level, the input data of the national energy sector model are adjusted with a new batch run of model calculations performed afterwards. In doing so, each subsequent iteration may be subject to changes

Table 5.2 The increase in investment risks under an increase in annual new capacity additions for nuclear power plants, %

Option	Capacity, MW			
	800	1100	1500	1700
Expensive gas	0–2	6–25	22–40	35–45
Cheap gas	1–10	16–36	32–45	42–55
Higher demand (10% higher than the reference value) for electrical energy when the gas is cheap	0–1	5–21	13–30	31–42

Note Calculated based on the stochastic optimization model (MISS-EL) for one of the Unified Energy System development scenarios in years 2020–2025. The lower bound was arrived at by assuming the normal probability distribution of the input data, the upper one corresponds to the interval uncertainty

in constraints on new capacity additions that possess unacceptably high investment risks properties. Directions and throughput capacities of cross-regional energy links may also be adjusted to mitigate the threat of a possible capacity shortage.

The totals of all calculations provide an overview of a possible (modeled) change over time for investment risks and the capacity shortage threat, its magnitude, and likelihood under assumed scenarios of economic development and under various hypotheses with respect to pricing policy adopted within the energy sector.

Table 5.2 makes use of a test example to demonstrate the dependency relation between the risks (as perceived by a potential investor) of further power plant capacity additions, the demand for the electric energy, and the probability density function used to model the uncertainty (including that of technical and economic performance indicators for various types of power plants). It shows that the risk of adding 800 MW under assumed conditions does not exceed 10%, while under the total addition of 1700 MW it increases up to 34–46% if the expensive gas assumption holds true or up to 24–55% if the gas price decreases by 25%. Obviously, the decrease in investment risks is facilitated by an increase in the demand for the electric energy and by more reliable input data projections.

The assessment of risks and the severity of a possible capacity deficit can be communicated by means of the indicators listed in Table 5.3.

The approaches employed to identify investment risks were covered above. Of most complexity is the assessment of possible damage due to a capacity deficit in the energy sector.

The long-lasting (ranging from one year to several years) and large-scale (starting from several million tons of coal equivalent or dozens billions kWh) shortage of a certain energy carrier leads to the relative decrease in production in the industries that act as energy and fuel consumers, as well as to the decrease in the demand for the products of the industries that serve as suppliers and are linked to the energy sector. In doing so, the chains of involved production links can be of considerable length. They get cut off in the case of substitution by imports, however, additional imports may require an increase in the exports of products that enjoy a sustainable demand on external markets, in order to preserve the national balance of payments.

Table 5.3 Indicators telling of the possible capacity deficit threat for regional systems and systems of individual industries of the energy sector

Indicator	Formula	Legend
The risk of an energy supply option (the shortage threat) [RS] The share of total capacity that possesses unacceptable high risk [CUR] Severity of the shortage threat [SST] Investment risk of the industry and the energy sector development option (the shortage threat) [RS] The share of total capacity that possesses unacceptable high risk [CUR] Severity of the shortage threat [SST]	Region [R] j $$\text{STR}_j = \sum_i r_i N_i / \sum_i N_i$$ $$\text{CUR}_j = \sum_i \bar{N}_i / \sum_i N_i$$ $$\text{SSTR}_j = \sum_i (l_i - \bar{3}_i)\bar{r}_i$$ The national economy [E], the energy sector $$\text{STE} = \sum_j \text{STR}_j \gamma_j$$ $$\text{CURE} =$$ $$\sum_j \sum_i \bar{N}_{ij} / \sum_j \sum_i N_{ij}$$ $$\text{SSTE}_1 = \sum_j \text{SSTR}_j \gamma_j$$ $$\text{SSTE}_2 = \Delta F \, \Delta \text{STE}$$	N_i—capacity additions, r_i—investment risks, $\bar{N}_i$—investment projects with unacceptable high risk, $\bar{r}_i$—average weighted risk of projects $\bar{N}_i$, l_i—damage due to a shortage, $\bar{3}_i$—costs to eliminate the shortage, γ_j—the share of total capacity provided by region j, ΔF and ΔSTE—changes in the objective function and risk values in the employed models after introducing constraints for $\bar{N}_i$.

Note Indicators are calculated for individual years of the projection time frame but account for the specific features (including the damages) of a given option (scenario) as it unfolds in time

Its additional production also requires energy resources. Damages due to the insufficient development of the energy industry may continue well after the shortage is eliminated, especially if the latter slows down the development of the capital assets forming industries and makes investment resources more scant.

All of the above contributes to the complexity of production links and forces us to account for their dynamic nature and ambiguity. The extent and intensity of the negative impact of an energy carriers shortage to a large extent depend on how its impact is distributed over individual industries and regions.

When making projections for the electrical power industry development, in a number of developed countries, the average specific damage due to a shortage of the electric energy is assumed to be about 10 US dollars per kilowatt-hour. The up-to-date assessment of the specific damages due to insufficient supply of electric energy for several regional energy systems (the Lenenergo, Orenburgenergo, and Rostovenergo electric utilities) presented in [27], assumes the value for this indicator that is significantly lower: 3–4 US dollars per kilowatt-hour (as per the 2010 ruble exchange rate). The reference guide for electric power system engineering [28] suggests to take the value of the specific damages due to the electric energy supply failure as equal to 1.5–4.5 US dollars per kilowatt-hour.

In [29], based on an analysis of reported data on the electric power consumption, GDP, gross added value, and electric energy efficiency in Russia in 2008 and 2010, the minimum specific damages to the economy of Russia due to failure of the electric power supply to consumers is estimated as equal to 5.95 US dollars per kilowatt-hour.

The above numbers are valid for the damages due to a short-time electric energy shortage. The damages due to a long-lasting shortage, that are caused by the energy sector capacity lagging behind its required development rates, are a way more devastating. Cross-industry models can be employed for their assessment. The first attempts at using the models of this kind were made as early as in 1983 by Gershenzon [30]. It follows from his calculations that the insufficient development of the fuel industry can lead to the decrease in the national income in a matter of 5 years by approximately 8 rubles for each ruble of the cost of the non-supplied fuel.

Currently, a system of models that includes the MIDL multisectoral model of the economy is employed at the Melentiev Energy Systems Institute of the Siberian Branch of the Russian Academy of Sciences to obtain such estimates. The estimates given in Fig. 5.2 and Table 5.4 for possible macroeconomic consequences of a capacity shortage have been obtained with the use of this tool.

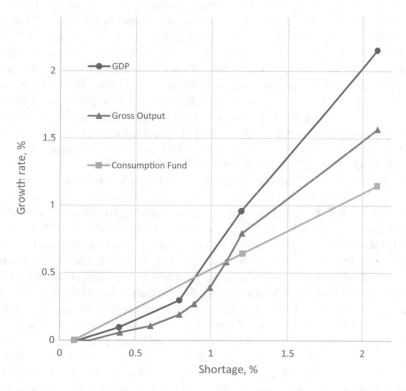

Fig. 5.2 The decrease in the values of macroeconomic performance indicators under a long-lasting gas shortage in the production sector (for the conditions averaged over Russia in 2010, with exports and imports unchanged). *Source* Calculated by the author

Table 5.4 Possible annual GDP decrease due to the 1% shortage of the energy sector industries capacity, %

Shortage	Electricity	Gas	Coal	Liquid fuels
Annual	0.15	0.15	0.2	0.3
Five-year	0.45	0.5	0.47	0.65

Note Relative to the reference case for expected conditions at the 2015 level.
Source Calculated by the author

5.3   National Security Threat Assessment with Respect to Unacceptably High Growth Rates of Price of Energy Carriers

Rough estimates of strategic threats to energy and economy security, including the price-induced ones, should receive due consideration from the Government in their social and economic projections [31]. Such projections provide averaged fuel price dynamics as well. At the initial stages of the energy sector projections development, these prices can be assumed to be acceptable for the consumer. They can serve as reference values when that determine probable regional market prices and threshold values of energy security indicators. Such indicators include the share of costs of production of energy-intensive products and of the private consumption that the cost of energy carriers accounts for. [1] The extent of expected threshold-crossing in some of the regions provides an estimate of the likelihood and the significance of the price-induced threat at the regional level.

An estimate of the likelihood of this threat can also be obtained by comparing projected market prices to the prices that ensure the minimum required production return rate and fuel and energy supply to consumers of a given region. Such assessment of a price-induced threat is aligned with the assessment of the possible capacity shortage threat in the fuel supply of a given region due to investment risks.

The indicators that are telling of price risks may include the following ratios as well: the ratio of market prices for energy carriers to levelized costs; the ratio of regional prices to the prices averaged over Russia; the ratio of fuel prices in Russia to the prices of the same fuel in net importer countries; the ratio of price growth rates for energy carriers (with inflation factored out) to GDP growth rates. The ratio of costs of the total energy consumed at the national scale to the GDP can also serve as an energy security indicator. In [32], the authors go as far as to provide a threshold value for this indicator, while pointing out that as soon as the share of such costs exceeds 10% of the GDP, economies of many countries will undergo through crisis developments. In Russia, the above ratio in 2013 was equal to about 8.5%.

The consistency (both across various countries and over time) of the ratio (share) of the energy consumption to the GDP (8–11%) or to the gross output (4–5%) in the

---

[1] As thresholds for the above indicators, one can benchmark the values they take abroad. In Japan and major European countries, the gas cost makes up 5–8%, while electric energy costs are 2–5% of total production costs [33].

long term is construed in [34] as one of the laws of the transformation of the energy backbone of a civilization. The author (Igor A. Bashmakov) points out the following: "When there is a significant 'crossing' of the upper bound (the threshold) of the solvency limits due to the energy costs growth, the unaffordability of energy slows down the economic growth. Such threshold values that are telling of the consumer ability to pay for energy resources are applicable for individual sectors as well: in the industrial sector (10–15% of the total revenue), the transportation sector (2–4% of the total revenue), the residential sector (2–4% of the total revenue)" [34, p. 21].

The battery of indicators for monitoring the dependence of the economy and the social sector on the changes in the energy sector products costs may include price elasticity of demand for fuel and energy on regional markets, but the key role is to be played by the indicators of the decrease in the cardinal value or growth rates of macroeconomic performance indicators (GDP, final consumption, investment in capital assets) per each percent of the increase in the price of fuel and electric energy for the consumer. Quantitative assessment of the effect of changes in fuel and energy costs on the dynamics of macroeconomic performance indicators requires due consideration of basic interactions between the energy industry and the economy (Fig. 5.3).

Econometric models that make use of statistical data do not satisfy the above requirements as they provide estimates that are way too coarse and ill-suited for projections.[2] A more justified approach would be to apply input–output price models. They enable us to estimate the effect of the propagation of price signals over the entire economy but fail to determine the change in the demand for energy carriers in response to the changes in their price.

By employing a system of economic and mathematical models one contributes to improving the validity of projection estimates of possible macroeconomic conse- quences of the change in the prices for fuel and energy resources. In the USA, one of such systems is the system of models developed by Data Resources Inc. (DRI). It includes the following: a dynamic model of economic growth based on neoclassical principles of total equilibrium, an optimization model of energy supply, and a static input–output model. The feedback between the energy system and the economy is channeled through fuel and energy prices (shadow prices) as obtained by the energy supply structure optimization.

In the studies that were carried out based on this model system in 1994 by the Electric Power Research Institute (ERPI) [36], the authors estimated a possible effect on the US economy caused by the inevitable increase in the cost of energy carriers due to an attempt to lower CO_2 emissions by means of introducing so called carbon taxes starting from 1997. Three carbon taxes scenarios were studied in juxtaposition with the reference case of the US energy and economy development: they assumed 50, 100, and 200 dollars per ton of carbon respectively. The introduction of the

---

[2]Professor Dr. A. I. Kuzovkin, at the meeting held by the NP "NTS ES" board (April 2013) [35], presented the following results of econometric calculations that appear controversial in terms of their significance: the growth of the real (i.e. that above the inflation rate) price of electric energy by 1% leads to the decrease in the GDP by 0.06–0.2%.

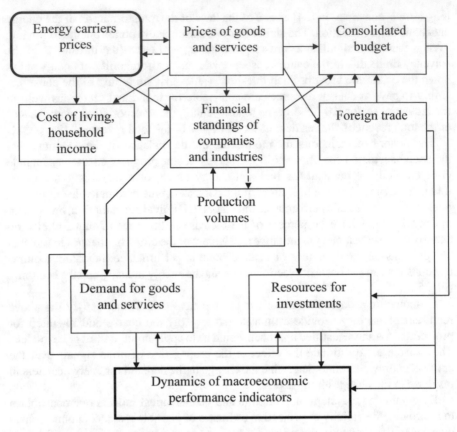

Fig. 5.3 The simplified flowchart of the impact of changes in the cost of energy carriers on the macroeconomic performance indicators

above tax would lead to a significant increase in the price of the fuel (and that of coal in particular) and the electric energy (see Table 5.5) and, despite a decrease in the demand for energy carriers (by 6–12% on average given the tax that would amount to 100 US dollars per ton), would cause a price surge: by 3.3% in the overall economy, and by 7.4% in the industrial sector (by year 2000, under the 100 US dollars per ton case). The resulting negative effect on the US economy due to an increase in energy carriers prices in the event of the introduction of carbon taxes is demonstrated in Table 5.6.

In Russia, a diverse range of model systems is also undergoing the ever more extensive development and find their way into delineating price relationships between energy systems and the economy [37–39]. One of the systems of this kind (INTEK) has been developed by the author at the Melentiev Energy Systems Institute of the Siberian Branch of the Russian Academy of Sciences (see Fig. 5.4).

The INTEK calculations are performed as per the procedure outlined below:

Table 5.5 Expected increase in energy carriers prices in the USA due to the introduction of the carbon tax (100 US dollars per ton), %

Economic sector	Energy carrier	Year 2000	Year 2010
Residential sector	Gas	23.4	18.3
	Electricity	27.1	23.9
Commercial sector	Gas	27.6	21.4
	Electricity	27.2	25.2
Industrial sector	Gas	42.7	39.9
	Coal	167.0	149.0
	Electricity	37.2	34.2
Transportation industry	Gasoline	18.4	15.6
	Diesel fuel	22.1	17.8
Electric power industry	Gas	52.3	36.8
	Coal	190.0	166.0

Source [36]

Table 5.6 Anticipated effect of the carbon tax on the US economy (100 US dollars per ton), %

Performance indicators	Year 2000	Year 2010
GDP	– 0.7	– 2.3
Consumption of goods and services	– 0.3	– 1.9
Capital expenditures in the industrial sector	– 2.5	– 4.6
Capital expenditures in the non-industrial sector	– 2.3	– 3.2
Exports	– 0.4	– 1.9
Imports	– 1.2	– 2.9
Inflation (GDP deflator)	2.3	3.9

Source [36]

1. The macroeconomic model (MIDL), the energy consumption model, and the MAKROTEK model are set up so as to be aligned with the reference case of the energy and economy development that assumes certain price dynamics for energy carriers. Assumptions with respect to scenarios of such dynamics are made.
2. A possible change in prices in response to the above for individual industries of the industrial sector is calculated (the INFLATION model). To this end, production volumes and material intensity ratios at this (initial) calculation stage are assumed to be the same as in the reference case.
3. A possible effect of the changes in prices on total revenues of industries, the population and the budget, and thus on the changes in boundary levels of final consumption of goods and services and on the resources available for capital expenditures (the OGRAN model).
4. Given the above changes the constraints of the MIDL macroeconomic model are adjusted with MIDL calculations performed anew afterwards.

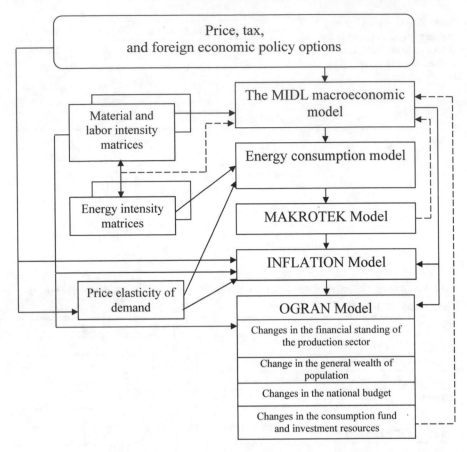

Fig. 5.4 A system of models for approximate estimates of macroeconomic consequences of a price policy adopted in the energy industry

5. The new calculation results obtained with the MIDL macroeconomic model get juxtaposed with the previous ones. In case they differ substantially, material intensity ratios (matrix A) are revised (given the changes in the ratio of existing to newly added production capacity) as well as labor and energy intensity ratios.
6. The corresponding changes get incorporated into the INFLATION and OGRAN models and a new iteration of calculations based on these models is then initiated.

Results of calculations performed in 2010 following the outlined procedure as applied to the best-case scenario of the economic development (with average annual GDP growth rates of 5.6%) are presented in Tables 5.7 and 5.8.

The obtained quantitative estimates, tentative as they are, elucidate the nature and the significance of the negative effect on the economy due to the increase in the electricity price. This effect depends on the price increase rate in a non-linear way and decreases over time. The latter tendency is attributable to a hypothetical decrease of

Table 5.7 Minimum price increase in individual economy sectors given an increase in the electricity price, %

Industries	Growth in electricity tariffs, %					
	20%		50%		100%	
	Year 2010	2030	Year 2010	2030	Year 2010	2030
Machine industry and metal industry	0.07	0	0.52	0	1.27	0.25
Construction industry	0.02	0	0.56	0.12	1.46	0.65
Electric power industry	20	20	50	50	100	100
Oil production	0	0	0.6	0.6	1.99	1.92
Oil refinery	0.16	0	1.36	0.96	3.52	2.73
Gas industry	0	0	0	0	1.13	0.74
Coal industry	0.67	0.32	2.4	1.64	5.31	3.88
Iron and steel industry	0.77	0.23	2.57	1.24	5.6	2.94
Non-ferrous metal industry	1.61	0.71	4.31	2.05	8.82	4.38
Chemical industry and petrochemical industry	0.88	0.41	2.44	1.22	5.1	2.64
Forest industry, wood processing industry, and pulp and paper industry	0.27	0	1.18	0.41	2.72	1.19
Construction materials industry	0.63	0.19	2.16	1.1	4.76	2.71

Source Model calculations by the author

Table 5.8 Changes in macroeconomic performance indicators given an increase in the electric energy price, %

Performance indicators	Electricity tariffs growth rate					
	20%		50%		100%	
	Year 2010	2030	Year 2010	2030	Year 2010	2030
Inflation rate	0.7	0.3	2.2	0.95	4.8	2.3
Cost of living	0.8	0.7	2.4	1.9	5.2	4.2
GDP	−2.0	−1.6	−3.6	−3.1	−6.3	−5.5
Profits	−3.5	−1.9	−4.7	−2.8	−6.5	−3.9
Final consumption of goods and services	−2.4	−1.6	−4.5	−3.2	−8.1	−5.9
Capital expenditures	−1.9	−1.1	−2.6	−1.5	−3.4	−2.1

Source Model calculations by the author

the energy intensity in the majority of industries as well as by the structural changes taking place in the economy.

One arrives at similar conclusions based on the results of the modeled analysis of the dependency relation between the GDP and the increase in the fuel and energy price that is carried out at the Melentiev Energy Systems Institute of the Siberian Branch of the Russian Academy of Sciences [38, 39]. It follows from the results of their calculations that the current electric energy price elasticity of the GDP growth rates amounts to -0.16, while this sensitivity is expected to decrease to -0.12 in medium and long terms. The same indicator when applied to gas prices takes the value of -0.038 and is expected to decrease to -0.022 in the future.

The authors of the Federal Energy Committee analytical report [40] claim that the values the elasticity of the GDP, as based on the natural monopolies prices, falls within the -0.15 to -0.20 range, that is to say that given the relative price growth for energy by 10% the GDP growth slows down by $1.2-2\%$ annually.

The process of adjustment of the economy to an anticipated price surge in the energy sector takes time for the changes in the sectoral make-up, technologies as well as the lifestyle to happen. That is why approaches to the comprehensive assessment of this price-induced threat prove an indispensable subject of projections of the range of the acceptable energy sector development. It is one of the major problems that it to be solved when developing an energy strategy and policy.

The quantitative assessment of possible damage to the economy and the social sector brought about by an excessive increase in energy carriers prices or a capacity shortage in the energy sector is the most challenging of all components of a projection. However, without such estimates one can hardly claim any valid indicator values that are when exceeded can lead to a crisis in the energy sector as well as the national or regional economy.

The proposed approaches to an approximate assessment of possible consequences of the materialized threats require a variety of projections of fuel and electric energy prices as well as a significant development to be undergone not only by the models themselves but also by the very accounting for complex interactions between various strategic threats.

References

1. Ang BW, Choong WL, Ng TS (2015) Energy security: definitions, dimensions and indexes. Renew Sustain Energy Rev V(42):1077–1093
2. IEA (2011) Measuring short-term energy security. p 15. URL: https://www.yumpu.com/en/document/view/19621056/measuring-short-term-energy-security-iea
3. Sovacool BK, Mukherjee I (2011) Conceptualizing and measuring energy security: a synthesized approach. Energy V(36):5343–5355
4. Bushuev VV, Voropai NI, Mastepanov AM et al (1998) Russia's energy security. Nauka, Novosibirsk, p 302 (In Russian)
5. Pyatkova NI, Rabchuk VI, Senderov SM et al (2011) Russia's energy security: outstanding issues and available solutions. Voropai NI, Tcheltsov MB (eds). Nauka, Novosibirsk, p 198 (In Russian)

6. Senderov SM (ed) (2017) Ensuring Russia's energy security and deciding on priorities. Nauka, Novosibirsk, p 116 (In Russian)
7. Energy Security Doctrine of the Russian Federation (Pr-3167 dated November 29, 2012). URL: http://rosenergo.gov.ru/energy_security/normativnopravovaya_informatsiya (In Russian)
8. Senderov SM, Pyatkova NI, Rabchuk VI et al (2014) An approach to energy security performance monitoring at the regional level in Russia. Melentiev Energy Systems Institute, Siberian Branch of the Russian Academy of Sciences, Irkutsk, p 146 (In Russian)
9. Senderov SM, Pyatkova NI (2000) An application of the two-level research technique when addressing energy security issues. Izvestiia RAN. Energetika 6:31–39 (In Russian)
10. Smirnova EM, Senderov SM (2018) Energy security of Russia's regions: the current state and changing patterns over the last six years. Energeticheskaya politika 1:16–23 (In Russian)
11. Senderov SM, Rabchuk VI (2015) Challenges faced in the analysis of Russia's energy security at the federal level: approaches to the assessment of threshold and current values of top priority indicators. Izvestiia RAN. Energetika 5:3–15 (In Russian)
12. Senderov SM, Rabchuk VI, Pyatkova NI (2009) An analysis of complying with energy security requirements when pursuing various directions of the national energy sector development to the year 2020. Izvestiia RAN. Energetika 5:17–32 (In Russian)
13. International Energy Agency (2007) Energy security and climate policy: assessing interactions. OECD/IEA, Paris, p 156
14. Kruyt B et al (2009) Indicators for energy security. Energy Policy 37(6):2166–2181
15. Sovacool BK (2012) The Routledge handbook of energy security. Routledge, p 455
16. APERC (2007) A quest for energy security in the 21st century: resources and constraints. APERC, IEEJ, p 100
17. Cherp A, Jewell J (2014) The concept of energy security: beyond the four As. Energy Policy V(75):415–421
18. WEC (2017) World energy trilemma index: monitoring the sustainability of national energy systems, p 145. URL: https://www.worldenergy.org/wp-content/uploads/2017/11/Energy-Trilemma-Index-2017-Report.pdf
19. WEF (2017) Global energy architecture performance index (Report), p 32. URL: http://www3.weforum.org/docs/WEF_Energy_Architecture_Performance_Index_2017.pdf
20. Index of U.S. Energy Security Risk (2017 edition). Global Energy institute U.S. Chamber of Commerce, 2017, p 89
21. International index of energy security risk: assessing risk in a global energy market. Global Energy institute U.S. Chamber of Commerce, 2018, p 80
22. Yao L, Chang Y (2014) Energy security in China: a quantitative analysis and policy implications. Energy Policy V(67):595–604
23. Senchagov VK (2001) Economic security as the basis for ensuring Russia's national security. Voprosy ekonomiki 8:64–79 (In Russian)
24. Bezuglova MA (2011) The energy component of economic security as a factor of critical services and development of Russia. Yuridicheskyi institut, Saint Petersburg, p 198 (In Russian)
25. Ministry of Energy of the Russian Federation. Energy Strategy of the Russian Federation to 2035. Draft dated February 7, 2014
26. Bondarenko AN et al (2000) Adoption of the two-level research technique for solving energy security problems. Izvestiia RAN. Energetika 6:31–39 (In Russian)
27. Nepomnyashchyi VA (2010) Economic losses due to electricity disruption incidents. The publishing house of the Moscow Energy Institute, p 188 (In Russian)
28. Karapetian I, Faibisovich D, Shapiro I (2012) The book of reference for designing power grids. The ENAS Novations Center, Moscow, p 376 (In Russian)
29. Nepomnyashchyi VA (2012) Estimating losses due to disruptions of power supply to consumers. Akademiia energetiki 5:12–17 (In Russian)
30. Gershenzon MA (1983) Modeling of the dynamics of cross-industry links of the energy industry. Nauka, Novosibirsk, p 240 (In Russian)

31. Ministry of Economic Development of the Russian Federation. The projection of the long-term social and economic development of the Russian Federation to the year 2030 dated March 2013 [Electronic Publication]. Retrieved from: http://www.consultant.ru/document/cons_doc_LAW_144190/ (In Russian)
32. Tsibulsky V (2014) Erratic wanderings in search of a magic potion. Nezavisimaya gazeta. Energiya 2(85):9–10 (In Russian)
33. Vedomosti, No. 5, 2014
34. Bashmakov IA (2013) The development of comprehensive long-term programs of energy saving and better energy efficiency. Methodology and hands-on experience: the authorized summary of a dissertation submitted in partial fulfillment of the requirements of Doktor Nauk in economics. Moscow, p 53 (In Russian)
35. Minutes of the meeting held by the board of NP "NTS ES" ['Scientific and technical council of the Unified energy system' not-for-profit partnership] (Moscow, April 2013)
36. Economic impacts of carbon taxes: detailed results. EPRI TR-104430-V2, Project 3441-01, Final Report, November 1994, p 320
37. Uzyakov MN (2004) The effect of prices of energy carriers on the dynamics of Russia's economy. Regiony i Federatsiia. Voprosy regulirovaniia TEK 2004 1:14–29 (In Russian)
38. Makarov AA, Malakhov VA, Shapot DV (2001) Macroeconomic consequences of an increase in energy carriers prices. TEK 2:51–52 (In Russian)
39. Malakhov VA (2012) Estimating a dependency of the GDP and the demand for energy carriers on an increase in the cost of fuel and energy. TEK Rossii 1:32–37 (In Russian)
40. The analytical report submitted to the Government of the Russian Federation "On the effect of prices and tariffs of natural monopolies on Russia's economy and on the measures of the state tariff policy with respect to natural monopolies for the year 2002 and in the medium run". Vestnik FEK Rossii 6 (In Russian)

Conclusion

The energy sector development unfolds under the growing complexity of its external and internal interrelationships, drastic changes in the nature and sectoral make-up of the economy, globalization, and the accelerating rate of scientific and technological advances. These and other factors contribute to the uncertainty of the future and urge to be on the lookout for new approaches that would lower the uncertainty and increase the validity of long-term projections.

Our analysis of projections of energy systems development published in Russia and abroad prove the non-linear nature of the escalation of the uncertainty range as the projection time frame extends further into the future. At the same time, this makes the accuracy requirements, that are to be met by projections, much less stringent. Our calculations suggest that the dependency of the value of investment projects for power plants of various types on changes in the demand for electric energy, fuel prices, and the discount rate is significant only within the first 15 years of their expected lifespan.

Accordingly, as the time frame increases projections should prioritize all the more not numeric estimates of energy carriers production and consumption volumes and even less so the guesstimates of their prospective values, but the identification of new trends and threats, the justification of the limits put on possible and feasible energy sector development options, the assessment of the significance of strategic-level threats and challenges, and the identification of their patterns as a function of changing external and internal conditions.

The methodology behind the modeling of the long-term energy sector development and the application practices adopted by those who employ a diverse range of mathematical models and the so named "model and information systems" should incorporate the principle of correspondence between the complexity and the level of granulation of research tools, on the one hand, and the intrinsic uncertainty of input data and accuracy requirements specific to the problem they deal with, on the other hand.

The principle of correspondence is fulfilled in our incremental approach to narrowing down the uncertainty range of conditions and results. This is achieved by iterative calculations performed on models of different hierarchical levels. A subset

© Springer Nature Switzerland AG 2020
Y. D. Kononov, *Long-term Modeled Projections of the Energy Sector*,
Springer Geophysics, https://doi.org/10.1007/978-3-030-30533-8

of models catering to a specific time frame of the projection is employed, which is followed by the harmonization of final values of performance indicators in time. To this end, at the initial stage one assumes the most extensive time frame (over 25–30 years) and the least number of hierarchical levels and models.

A major takeaway from the analysis at this stage is a projection of the dynamics of demand and prices in the energy sector with due consideration of a possible effect of anticipated new trends in scientific and technological advances as they manifest themselves in the energy carriers production and consumption, the national economy, and the quality of life. Doing away with such long-term projections of these performance indicators makes it difficult to arrive at evidence-based understanding of the efficiency and the riskiness of the energy sector and large-scale projects development options that are identified at the stages that focus on the development of short-term and medium-term projections.

The incremental approach to projects that we advocate is based on the retrograde movement from the more remote future to the near future. However, it does not preclude a subsequent reverse iteration of the projection: i.e. a revision of long-term projections so that they could accommodate the results obtained by a more granular analysis of a shorter time frame. Iterative calculations carried out at each stage that target a specific segment of the overall time frame (as performed both in top-down and bottom-up fashions) make it possible to account for the features unique to the development of systems (their opportunities and demands) of varying hierarchical levels that constitute the national energy system. To this end, it is practical for projections that cover 15–20 years ahead to account for possible responses to projected price and demand changes by potential investors.

Contingency calculations carried out at each stage assume scenarios of the economic and global energy markets development that are of varying levels of detail. Such calculations define the range of all possible values envisaged by a given projection of the energy sector development and help identify invariant solutions and contingent areas within it (under various criteria and conditions). The proposed approach to the exploration of this zone is based on the risk assessment of the facilities that make it up. Here, the risk value is derived from the inverse of the frequency with which the above facilities make it to the set of balanced out optimal solutions. The less frequently a given facility appears in such solutions the higher its investment risks are. The large number of tests under various combinations of input data, which is a desideratum of an analysis of this kind, can be obtained by blending optimization with Monte Carlo simulation within a single model.

An important role in the study of projections is played by singling out the problems that are of the utmost significance within each segment of the overall time frame and then solving them with appropriate methods. In this book, we focus on the approaches to tackle the following two of them: the problem of long-term projections of the possible state of regional markets (as defined by prices and demand), and the problem of the quantitative assessment of possible barriers that hinder the development of the energy sector on a par with strategic-level energy security threats under assumed scenarios of the economic development.

To solve the first of the two, we simulate the competition of energy carriers and their suppliers in a given region under incomplete information. To assess a possible response of consumers and investors to a change in the energy carriers cost we use original models that blend optimization with methods of statistical testing. To this end, we account for price elasticity of the fuel and energy demand. Our calculations provide evidence for a significant variation over regions and the dependency on the nature of the input data uncertainty.

The values of the price elasticity of demand for individual regions may serve as an energy security indicator: the less the elasticity is the less the possibility for mutual substitution of energy carriers becomes and the more urgent the energy undersupply threat proves in the case of, for example, a gas supply shortage.

Among the barriers that threaten the required energy sector development a prominent role is played by inertia. The identified non-linear dependency of its indicators on the growth rate and the make-up of the energy sector can be rationalized on the basis of high capital and material intensity of the constituent industries and a large share of capital expenditures for linked industries and manufacturing facilities. The importance of these indirect costs increases with higher growth rates of the fuel and energy production but can be lowered by means of higher import levels for equipment and materials. Available methods of quantitative assessment of such inertia indicators as the time and the scale of required look-ahead development of the industries linked to the energy sector and of corresponding capital expenditures can contribute to the improvement of approaches to comparative ranking of the energy sector development options based on the feasibility criterion. Evaluation of risk in individual capital investment projects should serve as a vital part in the process of arriving at such a comprehensive criterion.

The set of all strategic-level energy security threats can be divided into two major groups: the threat of new capacity additions lagging behind the most likely demand scenario and the threat of an excessive increase in the cost of energy carriers. The approaches we champion in this book and elsewhere to handle an approximate assessment of these interrelated threats provide for the identification of the timing and the probability of their occurrence on a par with their severity: the magnitude of possible losses due to the threat event materializing and the costs to mitigate them.

We propose a two-level (that is, encompassing both the national and the regional levels) approach to the assessment of the probability and the severity of the capacity shortage threats. To this end, the problem is treated as an analysis of possible risks of the energy and fuel supply of a given territorial entity under uncertainty.

To obtain an approximate estimate of likely macroeconomic consequences induced by changes (relative to the reference case) in the fuel and electric energy prices one can employ the system of economic and mathematical models (INTEK) developed by the author.

The assessment of strategic threats has to be reflected in the values of energy security indicators. Ideas as to how to extend their composition as well as the methods to calculate them are presented in this book as well.

Obviously, the proposed methods of increasing the validity of long-term projections of the energy sector developments are subject to discussion and call for further research efforts. It is also clear that the list of the problems that are to be treated as part of the projections development goes beyond the scope covered in the book. It depends on a given time frame and the required lead time of decisions to be made.

Of top concerns that, as of today, still lack satisfying solutions one can highlight the following ones:

- Assessment of the possibility and conditions to adjust to a new scenario of the economic development and the state of world energy markets, as well as to the shift from one pathway of the energy sector development to another (in terms of required resources and the time it would take).
- Approaches to assessment and harmonization of the results of projections that are obtained by solving optimization problems at different hierarchical levels based on different criteria of economic efficiency (each appropriate at the levels of individual businesses, industries, and the national economy respectively) as well as the criteria of flexibility (the ability to accommodate changes), reliability, and security.
- Numeric estimates of threshold values of energy security, and national security indicators with due consideration of their functional dependency on the scenarios of social and economic national developments, on projections of scientific and technological advances in production and consumption of energy carriers, and, finally, on a given time frame and other contributing factors.
- The correspondence between employed methods and models of the input data uncertainty, on the one side, and required performance of projections, on the other side, as applied to various time frames. Identification of the minimum required level of detail appropriate for solving a given problem of the adequate representation of elements and links within a given system.

It appears that the above 'to do' list can also be extended to cover the identification of the pitfalls and the root causes of energy and economic systems getting trapped in a zone of instability and bifurcations where even the most negligible of changes in conditions is able to invoke a crisis or an emergency. Arriving at a solution of the latter problem can be facilitated by a dialogue with the cross-disciplinary field of synergetics that studies the laws of evolutionary processes, sustainability, and self-organizing of open non-linear dynamic systems.

Printed in the United States
by Baker & Taylor Publisher Services